THE COLD LIGHT OF DAWN

The discovery in 1987 of a supernova brought to world attention the excellence of Canadian astronomers. As Richard Jarrell explains in this book, the path to excellence has been a long one. Although astronomy has been practised in this country from the earliest days of exploration, its professional status has slowly evolved in much the same way as has the nation itself.

In the period of exploration and early settlement, the practical needs of navigators and surveyors were foremost. Astronomical practitioners – for many used astronomy but few were professional or even amateur astronomers – came from elsewhere. Only when Canada was a settled colony, halfway through the nineteenth century, did its own scientific needs emerge. By the century's end Canadian astronomy, socially and institutionally unique and independent, had been established: astronomers born and trained in Canada worked in their own organized and funded institutions.

In the twentieth century the story is dominated by the Dominion Observatory, the Dominion Astrophysical Observatory, and the David Dunlap Observatory, and, in higher education, the University of Toronto. The federal government remained the biggest actor, in employment and funding, first through the observatories, then the National Research Council. The expansion of universities greatly broadened the scope of Canadian astronomy, while the Royal Astronomical Society of Canada, local clubs, literature, planetariums, and museums kept the public informed. By the 1960s Canadian astronomy, though small in size, was as sophisticated as any in the world.

RICHARD A. JARRELL is Associate Professor of Natural Science at Atkinson College, York University. He is co-founder of the Canadian Science and Technology Historical Association and founding editor of *Scientia Canadensis*.

RICHARD A. JARRELL

THE COLD LIGHT OF DAWN:
A History of Canadian Astronomy

UNIVERSITY OF TORONTO PRESS

Toronto Buffalo London

© University of Toronto Press 1988
Toronto Buffalo London
Reprinted in paperback 2017

ISBN 978-0-8020-2653-8 (cloth)
ISBN 978-1-4875-9202-8 (paper)

Canadian Cataloguing in Publication Data

Jarrell, Richard A., 1946–
The cold light of dawn

Bibliography: p.
Includes index.
ISBN 978-0-8020-2653-8 (bound)   ISBN 978-1-4875-9202-8 (pbk.)

1. Astronomy – Canada – History. 2. Astronomical
observatories – Canada – History. I. Title.

QB33.C3J37 1988   520'.971   C88-093664-9

Cover illustration:
David Dunlap Observatory, at dawn
(photo: D'Arcy R.G. Jarrell)

This book has been published
with the help of a grant from the
Canadian Federation for the Humanities,
using funds provided by the
Social Sciences and Humanities Research Council
of Canada.

For my parents

Contents

Preface

This is the first general history of Canadian astronomy. It is my intention not to provide an exhaustive account of the subject but rather to describe the main points of interest in the evolution of Canadian astronomy, along with a historical framework to help explain – insofar as any historian can explain past events – the rise of astronomical science in Canada.

There is a growing literature on this subject; two monographs that have helped delineate the scope of the present work are Don W. Thomson's *Men and Meridians* (Ottawa, 1966–9), a monumental recounting of surveying in Canada, and Malcolm Thomson's *The Beginning of the Long Dash* (Toronto 1978). Don Thomson's subject overlaps often with astronomy, since, until the beginning of the twentieth century, most of Canadian astronomy dealt with longitude and latitude determinations. Therefore, this work bypasses much of the history of exploration and surveying. Malcolm Thomson has provided a detailed account of timekeeping in Canada, and the reader will be directed to his work for a fuller account.

This study focuses on the social and institutional history of Canadian astronomy. While technical details are essential at many points, this is not intended to be technical history. Indeed, there is little in Canadian astronomy from a technical point of view that is original until the twentieth century, but the groundwork for the contributions of contemporary Canadian astronomers was laid in the last century, particularly after 1840, and those roots are worth examining.

The reader will notice differences in style and emphasis for the periods before and after 1905. The history of this subject before the present century is more compact and more easily approached. There were few astronomers and few institutions, and the archival material that survives can all be surveyed. Twentieth-century science is a different matter: much of its subject-matter is still evolving rapidly. The number of scientists is much larger, growing from a half-dozen astronomers in 1860 to approximately 300 by 1980; much of the archival material is only partially open to the public; and a great deal of the correspondence is not available at all. Because most of the Canadian astronomers who ever lived are now alive and working, interviews with them or with the recently retired are often extremely illuminating but frequently – for very good reasons – consciously or unconsciously edited by the subjects. The historian simply cannot always discover what he or she wishes to know and must accept this as a fact of the craft. For the years to 1905, I have endeavoured to offer a reasonably complete account of the rise of astronomy in Canada. For the period after 1905, I have selected what I consider the most important advances made in Canada and the most significant changes in the social and institutional evolution of the science. I see this work as a summing up of several decades of historical work and hope that it inspires further investigation.

This work took shape over ten years, and during that time it was my pleasure to exchange information with many scientists, historians, and archivists. I would like to thank those who graciously allowed me to interview them and tape our conversations: the late Dr C.S. Beals, Prof A.V. Douglas, Dr J.A. Pearce, Mr Malcolm Thomson, and the late Prof J.F. Heard. These conversations not only provided basic information but also helped to establish a 'reality' for past events captured in published reminiscences. Dr K.O. Wright and Dr Harvey Richardson kindly allowed me to read the Plaskett correspondence at the Dominion Astrophysical Observatory. Dr Jack Locke and Prof J.E. Kennedy were helpful on several points of inquiry. Mr Arthur Covington's friendship has provided me with a steady stream of information, and Dr Helen Hogg's kind interest in my work over the years has been beneficial. Valuable information and support from Mr Phil Mozel and Prof Roy Bishop were appreciated. Yves Gingras, of the Université de Montréal, provided the information necessary for appendix C.

The Royal Astronomical Society of Canada provided much assistance: Mrs Marie Fidler and Ms Rosemary Freeman were always helpful, Mr Peter Broughton gave assistance and encouragement, and Ms Nora Hague of the Montreal Centre provided timely information. I owe a great debt to archivists across the country and in the United States: to Ms Sandra Guillaume, formerly at the McGill and University of Toronto archives, Mr Harold Naugler of the Public Archives of Canada, Mrs Mary Lea Shane of the Lick Observatory Archives, Mrs Mary Boone of the University of New Brunswick Archives, Mr Clark Elliott of the Harvard University

Archives, Mr Frank Zabrosky of the University of Pittsburgh Archives, and the archivists at Queen's University, Université Laval, and the Séminaire de Québec.

In addition, I must record the many kindnesses and aids from Dr Henry C King, former director of the McLaughlin Planetarium, and Prof Bruce Sinclair, of the Institute for History and Philosophy of Science and Technology of the University of Toronto. Special thanks are due to Mrs Sylvia Williams and Mrs Rita Marinucci, Atkinson College, York University, who steadfastly typed the drafts of this work. Finally, I am grateful to the late Mrs Mildred Mafit for proofreading various versions and most of all to my wife, Marti, who not only encouraged the work but also tried to reduce the number of linguistic infelicities in the manuscript; those that remain are due entirely to my own obstinacy.

Financial assistance for the initial research for this book was provided by the Canada Council, which is gratefully acknowledged.

THE COLD LIGHT OF DAWN

Introduction

Astronomy, it is often averred, is the oldest science. It is certainly the earliest science practised in what is now Canada, for knowledge of the night skies was essential for John Cabot's sailors who reached Newfoundland in 1497. Astronomy is still a significant, if small, part of the scientific effort of modern Canada. Like all sciences, astronomy attracts its practitioners for two reasons: it helps to solve practical problems, and it answers a basic desire for knowledge of the universe beyond the earth. In the nearly half-millennium from Cabot's voyage to our own time, the nature of the science of astronomy has changed profoundly. The majority of contemporary astronomers are involved in the intellectual side rather than the practical. Now a most sophisticated physical science, astronomy exhibits few obvious connections with the practical needs of everyday life in Canada. In that sense, it is peripheral; if all astronomical activity in Canada ceased tomorrow, almost no one would notice and no obvious economic effect would be felt. We must see modern astronomy not as a necessary adjunct to an industrialized nation but rather as an important cultural activity that is a measure of the level of civilization of a country. We support applied chemistry because we desire its economic benefits, but we finance the pursuit of astrophysics because as civilized, educated people, we seek knowledge of our physical environment for its intrinsic intellectual value.

The role played by astronomy in the evolution of Canada has itself changed over the course of five hundred years. Two basic transitions are described in this study. The first was the shift from astronomy *in* Canada to Canadian astronomy, from non-Canadians working for their own purposes to Canadians furthering their own needs. This transition reflects the evolution of Canada itself. Once Canadian astronomy emerged, it metamorphosed from almost purely practical astronomy to almost purely intellectual astronomy. This second transition takes place much later than the first, for it is a reflection not only of Canadian national evolution but also of that of the science itself.

The evolution of Canada from a seasonal outpost for fishermen to a thoroughly modern industrial nation of some twenty-five million people was a developmental process of much complexity. There is no agreement among historians about the nature of this development. Some see distinct phases in economic terms, others the interaction between Canada and its imperial masters, as in those frameworks characterized as 'metropolis and colony' or 'centre and periphery.' Still others look to the dynamic tensions between classes, or between regions and centre, English and French, Canadians and Americans.[1] All these ideas are plausible, but some basic requirements have to be satisfied before scientific activity can take root in a new country.[2]

Science is an important ingredient in the exploration and exploitation of the physical environment, and astronomy, like geology or natural history, was necessary from the beginning. So vast is Canada that the exploration, to say nothing of the exploitation, of natural resources is still incomplete after four centuries of settlement. For development, however, we must assume a permanent, growing population. Nor can we approach Canadian history in a linear fashion based only on population growth, for development was clearly a step-like process: the opening of the Canadian Arctic today has strong similarities to the opening of the west in the 1850s and 1860s, despite the technological differences. So long as the population of a colonial territory remains small and scattered, the colony must rely on outsiders – primarily the colonizing nation – for many of its economic, cultural, and scientific needs. The transition, in a general sense, from science in Canada to Canadian science required a large enough population, with sufficient economic and political power to displace the control and influence of the colonizing nation. This requirement was fulfilled only during the second half of the nineteenth century,[3] and the steps in the transition will be outlined in chapters 2 and 3.

The minimum population necessary for development is difficult to establish because of other factors. Density of population is more important than mere totals, making urbanization an important element. A rural economy cannot, by itself, engender much economic or social development. Towns attract people with capital, skills, and education. The slow shift from the primary economic activities of early Canada, such as fishing, fur trading, lumbering, or farming – along with associated industries such as milling, tanning, brewing, and distilling – to manufacturing went hand in hand with the rise of towns. In Canada, the move to secondary industry took place also in the second half of the last century. The growth of science required educated people and a variety of perceived needs, money, and institutions, all of which require urbanization and, increasingly, industrialization.

If we follow the growth of science in Canada, we see quickening of activity and an increasingly complex infrastructure during this same period.[4] The evolution of astronomy closely parallels that of the other sciences, except that its scale is always considerably smaller, for the needs of the Canadian community for astronomy were never as great as those for chemistry or the earth sciences. When Canada became a mature nation rather than a dependent colony, science could be cultivated for its intrinsic value more than for its practical contributions. There are examples of what we now call 'pure research' during the nineteenth century in Canada, but they are both rare and premature. Only in the twentieth century, and especially since the Second World War, has Canadian science come to resemble that of older, mature nations.

Before 1905, practical needs were paramount. Practicality is, thus, the thread of the evolution of Canadian astronomy until that time. As higher levels were achieved during the present century, the practical aspects of the science receded. Canadian development was unique; although Canada obviously shares many features with other colonial nations, such as the United States or Australia, its pathway diverged from theirs because its land, people, and social and political growth differed from theirs. It follows that Canadian science, and Canadian astronomy, also are unique. The difference is not so much in method – there was no uniquely Canadian way of determining longitude – but in structure. Canada's institutions, forms of education, choice of research topics, and means of funding all differ in some respects from those of other nations. A significant difference is the role of the state in the growth and support of science. While the state was often a patron of astronomy, providing, for example, the Observatoire de Paris, the Greenwich Observatory, and the United States Naval Observatory, the ubiquity of the state in Canada is striking. This role appeared early, as chapter 2 will show, but it is still very much present, as the reader will note in chapter 7.

The transition from colony to nation, the shift from practical to pure science, and the role of the state in producing a uniquely Canadian astronomy are the themes of this study. The social aspect of Canadian astronomy will, then, be as important as the technical for this work.

PART ONE
Astronomy in Canada
1534–1840

CHAPTER ONE

Colonial Astronomy

In purely temporal terms, the state of 'astronomy in Canada' occupies three-quarters of the history of Canada. From Canada's discovery in the last years of the fifteenth century until the second quarter of the seventeenth, there was no settlement of the country and only occasional exploration. A passage to the Orient, or riches such as gold and fish, were the attractions, not permanent European settlement. Even when New France began to receive settlers, Canada was not much of a success; in 1760, less than one-tenth of the people in the British colonies of America lived in what is now Canada. For two centuries after the 1640s, Canadians were few, scattered, controlled by foreign powers, and reliant upon their metropolitan centres for economic, cultural, and scientific sustenance. The Jesuit missionaries, an essential part of this story along with the explorers, came to convert the natives, not to practise science. They were, for the most part, Frenchmen, not Canadians, and the explorers, both French and English, were also foreigners. Exploitation, not the development of a viable colony, was the concern of the governments in Paris and London. Science was a tool for exploration and exploitation or, with the Jesuits, an intellectual pastime or subject of education. It was not part of an indigenous Canadian culture.

The European conquest of Canada began in 1497, when John Cabot arrived in Newfoundland, seeking a route to the East for his master, Henry VII of England. On his heels were others – English, Spanish, Portuguese, and French. Regardless of nationality, these explorers had in common a basic knowledge of astronomy. Although the late fifteenth and early sixteenth centuries were a lively era for astronomy, the great cosmographic questions of the day – the grand theories of Ptolemy, Copernicus, and Tycho Brahe – were not of interest to explorers and navigators, who saw the universe from a geocentric perspective. More valuable to them were better charts, ephemerides, instruments, timekeepers, and star catalogues – practical materials that would aid them in crossing the ocean and in determining their geographic position once they made landfall. These navigators and explorers were the early practitioners of astronomy in Canada, and their use of astronomy was not so much a scientific undertaking as an essential art.

The methods and instrumentation necessary for determining the latitude and longitude of a place, whether on land or at sea, were much the same for the French explorers along the St Lawrence and the British in the Arctic. Because British exploration of the Arctic dates back even further than French exploration of eastern Canada, we will look first at British instruments and methods. A selection of two works by explorers, one each from the seventeenth and nineteenth centuries, gives an idea of the level of advancement of British astronomical instruments and techniques. The first is *The Dangerous Voyage of Capt. Thomas James* (1633), an account of James's voyage into Hudson Bay in 1631–2; the second is *A Voyage of Discovery ... for the Purpose of Exploring Baffin's Bay* (1819), an account of Sir John Ross's expedition.

James's book is accompanied by two interesting appendices, one a list of instruments purchased for the voyage, including a 4-foot-radius pearwood quadrant, divided into minutes of arc; a similar 2-foot quadrant; and a wooden equilateral triangle. Along with these were 7-foot and 6-foot cross-staffs, three Jacob's staffs, a Gunter's staff, and two Davis backstaffs. The various staff instruments, which looked like crossbows, were simple ruled devices to give the angular distance between two objects or between an object and the horizon. They were naked-eye instruments, inclined to warp, poorly divided as a rule, and, even in the most experienced hand, rarely capable of providing accurate readings. The Davis back-staff allowed the navigator to 'shoot' the sun without looking at it directly, but the instrument was not a great improvement over the others. Along with an assortment of compasses, James had a meridian line with plumb bobs, a book of tables based on Gunter's calculations, and 'A Chest full of the best and choicest Mathematical Books, that could be got for Money in England; as likewise Master Hackluyt, and Master Purchas; and other Books of Journals and Histories.'[1]

The other appendix is a brief description of methods for determining longitude, written by Henry Gellibrand, who taught astronomy at Gresham College in London. Gellibrand explains how latitude and longitude were obtained at a time before chronometers were available. Latitude was calculated by one of two methods. First, latitude could be measured by finding the angle between the northern horizon and Polaris, after accounting for the fact that Polaris describes a small circle daily. Second, one could measure the sun's altitude at local noon. By consulting astronomical tables, the navigator could find the sun's declination – the angular distance above or below the celestial equator – for a given day, thus enabling him to easily calculate the latitude. Typically, however, he was plagued by a combination of problems, such as not being able to fix exactly local noon, instrumental errors, observer error, failure to include corrections for refraction, and, at sea, dip, unknown or imperfectly known before the mid-seventeenth century. At sea, miscalculating the latitude by as much as 15 minutes of arc, the equivalent of 15 nautical miles, was not uncommon.

Longitude presented a much more difficult problem. The idea that compass directions could aid the navigator, even if he knew the compass variation, is dismissed out of hand by Gellibrand. The two methods he approved were both employed by James in 1631. The first was the observation of the lunar eclipse of 29 October 1631. James was able to find the altitude above the horizon of the moon's upper limb at its emersion from the earth's shadow. Gellibrand, observing in London, found the time of the same event; computing from the tables of Phillip van Landsberg, which 'do much amuse the World with that lofty Title of Perpetuity,' he found that the difference in time between his observation and the time that James should have observed was $5^h 18^m$, giving a longitude for James's observation point, Charlton Island, of 79°30'w of London.[2] A second method involved the occultation – the passage of the moon in front of an object – of a star in the constellation of Corona Borealis on 23 June 1632. Since this had not been observed in London, Gellibrand had to calculate his time from Landsberg's tables, arriving at a longitude for Charlton Island of 78°30'w. The mean was 79°w. The modern value for Charlton Island is 79°15'w. As it turns out, this close agreement was probably due to a combination of errors cancelling each other. Most contemporary explorers were lucky to obtain values within even 5 degrees.[3] Such was the accuracy of the seventeenth century.

At sea one could not employ these methods easily, if at all, and navigators guessed at longitude by employing the log line and hourglass to estimate the daily traverse. Then, unwittingly, they compounded their error by using an incorrect value of the nautical mile, which was not accurately known until Jean Picard's work of 1671.

Moving forward to the early nineteenth century, we may contrast James's equipment and methods with those of Sir John Ross (1777–1856), whose first voyage to the Arctic was in 1818. On that voyage he took along seven chronometers, a pendulum clock, a transit, an altitude instrument and theodolite by Jones, and a transit and sextant by Dollond, along with astronomical tables. His officers were expected to take observations for latitude and longitude every day possible, the latter being checked by chronometer and by lunar distance observations. The clock errors were recorded, and normally the mean of the times of five chronometers was taken for longitude.

Ross's second voyage, 1829–33, had as sound a scientific basis as the first.[4] He took fourteen chronometers along, a Jones transit and theodolite, and five sextants. A telescope with $3\frac{5}{8}$-inch objective and 66-inch focal length, by the celebrated London optician Charles Tulley, completed the instrumentation. The telescope was for observing occultations. The interval between James's voyage and Ross's expeditions saw most of the important advances in celestial navigation. The sextant had replaced the larger, clumsy quadrant. It was a direct descendant of John Hadley's octant, introduced in 1732, which allowed the observer to bring the horizon and an object into the same plane with mirrors on moving arms. By Ross's time, these instruments had telescopic sights and vernier scales. Portable transit instruments, which had become the chief optical instruments for geographical measurements on land, were small telescopes that moved above a fixed east-west axis and had fixed, later movable, wires, which the object in view crossed. The transit instrument was invented by Roemer late in the seventeenth century, but it was not truly portable until late in the following century. The major breakthrough for longitude work was the chronometer, an accurate portable clock developed by John Harrison and perfected in the 1760s. For land-based operations, Huygens had developed the pendulum clock in 1660, but pendulum clocks were not both portable and accurate simultaneously.

Along with the improvements in clocks and instruments, the theory of celestial motions steadily increased in accuracy. The lunar tables of Tobias Mayer enabled Astronomer Royal Nevil Maskelyne to begin publication, in 1767, of the *Nautical Almanac*. Mayer's work was the latest improvement after the important work, especially in lunar theory, of Clairaut and Euler.

The explorer had to take Greenwich time with him by means of the chronometer, and Ross was being careful by taking several, since each one gained or lost time at a characteristic rate. Having several chronometers, and knowing their rates, the observer could have several independent checks on the time. Ross's officers typically employed the lunar distance method, using sextants to find the angular distance between the sun and moon, if both were visible, or the moon and selected stars. The distance between them for Greenwich at any given time could be computed from the *Nautical Almanac*. Because the observer had Greenwich time with him, he could then compute the time difference between his location and Greenwich, thus the longitude. Occultations of a few bright stars take place in any given year. The times of disappearance and reappearance of a star from behind the moon's limb were calculated in the *Nautical Almanac* for Greenwich. The time difference between the explorer's observation of the occultation and that predicted for Greenwich would give the longitude difference. Finding the latitude had not changed substantially since the time of Capt James, but defining the horizon was always difficult, especially at sea. The artificial horizon was introduced about 1750 and, by Ross's time, was an integral part of a sextant.

The great activity in exploration of the New World during the fifteenth and sixteenth centuries, particularly by the English and Spanish, was for the purposes of establishing new trade routes, building empires, and, especially for the Spanish, colonization. The French showed curiously little interest in such ventures. François I, in hopes of having a trade route to Asia, commissioned Jacques Cartier in 1534 to seek such a route. Two further trips followed, in 1535–6 and 1541. The French did not return officially until 1604, although French fishermen had worked the banks off Newfoundland for many years. It was Samuel de Champlain (1567–1635), one of Henri IV's geographers, who was chiefly responsible for bringing the French to America to stay. He was also the real starting-point for astronomy in Canada. The earliest explorers maintained records of the latitudes of the stations of their voyages inland but rarely bothered with longitude. Champlain, founder of Quebec and first governor of New France, regularly took solar altitudes with his astrolabe during his expeditions. On one such trip, in 1613, while travelling up the Ottawa River, he lost his instrument on a portage. The astrolabe, in a remarkable state of preservation, was discovered by a farmer in Renfrew County, Ontario, in 1867. Although Champlain does not mention the loss of the instrument in his journal, his latitude estimates before passing the portage are reasonably accurate; those afterwards are clearly estimated and largely in error.[5]

The determination of latitudes quickly became part of the orderly settlement of New France. With the appointment of Jean Bourdon (1601–1668) as engineer at Quebec and surveyor general of the colony in 1634, astronomy became an official duty. Bourdon was a talented surveyor and cartographer who made astronomical observations in connection with his survey of the rivershores around Quebec. He was the first known Frenchman in Canada to possess an astronomical telescope, a small Galilean refractor with a compass, gift of the Jesuits in 1646.[6] His most noteworthy successor in the seventeenth century was Jean Deshayes, who had trained with J.D. Cassini at the Paris Observatory. On his arrival in 1686, Deshayes observed a lunar eclipse and attempted to calculate the longitude of Quebec, which he found to be 72°13′ west of Paris, nearly 1°20′ in error. Observational error, probably mostly instrumental, and deficiencies in lunar tables accounted for the discrepancy. On Deshayes's death in 1706, his instruments and nearly forty-volume library of navigation and surveying works passed to the Jesuits' Collège de Québec.

Scientific interest in New France by the home government was never great but reached its zenith during the governorship of Galissonière.[7] Although he was recalled to France after only a brief stay, Galissonière maintained his interest in the colony. He made a case for a better survey of the coasts of Acadia, requiring professional astronomical work. Consequently, a voyage was commissioned, with Joseph-Bernard, marquis de Chabert (1724–1805), as astronomer.[8] This was undertaken in 1750–1; he ascertained the longitudes and latitudes of a number of places in Acadia (Nova Scotia), Île Royale (Cape Breton), and Newfoundland. His account,[9] after being screened by Galissonière and two notable astronomers of the Académie royale des sciences, Pierre Bouguer and P.C. Le Monnier, appeared in 1753. The professionalism and care that Chabert brought to this work are in striking contrast to the rough-and-ready work of the Jesuits. He had a quadrant with two telescopes, one fixed, and a micrometer eyepiece. Latitude observations took into account instrumental error and refraction. To improve accuracy, and to account for personal equation – the time lag between perception and recording of an event unique to each observer – his assistant repeated all observations.

For longitudes, Chabert employed the eclipses of Jupiter's satellites, occultations, and lunar distance methods, whenever possible employing all three. For occultations he had a 6.5-foot telescope. Just to double-check the longitude of Louisbourg, he compared his observation of an eclipse of a Jovian satellite with an observation by Fr Vendlingen, SJ, at the Madrid Observatory, Paris being overcast. Knowing the difference in longitude between Paris and Madrid, he could easily calculate the Paris-Louisbourg distance as $4^h8^m27^s$ difference. For other checks, he obtained observations from James Bradley, the astronomer royal of England.

Unfortunately, the best astronomy done in New France, the work of Chabert and of Joseph de Bonnécamps, came at the end of French control. There is a postscript to French astronomy in Canada: in 1768, J.D. Cassini visited Newfoundland and the remnant of French North America, St-Pierre, in order to test some new clocks made by Le Roy.[10] He was happy to leave the miserable weather of Canada behind for further tests in Morocco and Cadiz.

The Society of Jesus, founded in 1539 to combat the Reformation, was, by the early seventeenth century, the most powerful and ambitious and best-educated order in the Roman Catholic church. Its missions, and its astronomical activities, ranged from Canada to China. The Jesuit colleges set the standard for excellence in European education, with a course of studies including Aristotelian natural philosophy, mathematics, and geography, a subject that included astronomy. Both René Descartes and Marin Mersenne were educated in the Jesuit Collège de Laflèche, which supplied a number of missionaries to Canada. The Jesuits were educated men of the world, bringing to their missionary work not only religious zeal but also a profound interest in the lands and peoples where they worked. Astronomy was one of these interests.

The first appearance of the Jesuits in New France was in 1604, with their mission in Port-Royal, in what is now Nova Scotia. The earliest scientific report from them was penned by Biard in 1616. Throughout the seventeenth and eighteenth centuries, a stream of reports from the Jesuits was sent to France, not only recording the advancement of the missions, but also describing a great variety of scientific observations, including some in astronomy. Father Paul Le Jeune (1591–1664) attempted, as described in his 1632 report, to find the longitude of Quebec by a triangular solution.[11] He considered that his latitude was $46^2/_3°$, and that of Dieppe $49^2/_3°$. From sailors' estimates, he guessed the great circle distance to be 1,000 leagues. Solving by spherical triangle, and assuming a degree on a great circle to be 17.5 leagues – he did not know the difference in length of latitudinal and longitudinal degrees – he found the longitudinal difference to be $91°38'$, or about 19° too much.

In October 1633, Le Jeune observed an eclipse of the moon. He wrote his correspondent that, according to his almanac, the eclipse was to start at midnight in France but that he observed it at 6 p.m., confirming his longitude calculation of the previous year.[12] It is unlikely that Le Jeune had an accurate clock, if he had one at all, or that his almanac could have been very accurate by modern standards, but his reasoning was essentially correct.

Le Jeune was clearly interested in astronomy, for he recorded another lunar eclipse in 1635[13] and in 1637 gave a delightful account of teaching a native some astronomy and physics:

On the 10th of January Makheabichtichiou asked me many questions about the phenomena of nature, such as 'whence arose the Eclipse of the Moon?' When I told him that it was caused by the interposition of the earth between it and the sun, he replied that he could hardly believe that, 'Because,' said he, 'if this darkening of the moon were caused by the passage of the earth between it and the sun, since this passage often occurs, one would see the moon often Eclipsed, which does not happen.' I represented to him that the Sky being so large as it is, and the earth being so small, this interposition did not happen as frequently as he imagined; upon seeing it represented by moving a candle around a ball, he was very well satisfied. He asked me how it was that the Sky appeared to be sometimes red, sometimes another colour. I replied that the light, passing into the vapours or clouds, caused this diversity of colour according to the different qualities of the clouds in which it happened to be, and thereupon I showed him a prism. 'Thou dost not see,' I said to him, 'any colour in this glass; place it before thine eyes and thou wilt see it full of beautiful colours which will come from the light.' Having held it to his eyes and seeing a great variety of colours he exclaimed, 'You are Manitous, you Frenchmen; you know the Sky and the earth.'[14]

Note Le Jeune's Aristotelian explanation of colour; this episode took place thirty years before Newton's discovery of the basic properties of colour.

The Jesuit *Relation* of 1662–3 provides a good description of various aerial phenomena and the great earthquake of 1663. In the autumn of 1662, a brilliant fireball was seen over Quebec and Montreal, and fragments were observed emanating from it. Mock suns were seen in January 1663, and on 1 September a solar eclipse occurred.[15] The great earthquake occurred on 5 February 1663; the writer notes: 'The atmosphere was not without its disturbances, we saw specters and fiery phantoms bearing torches in their hands. Pikes and lances of fire were seen, waving in the air, and burning brands darting down on our houses – without, however, doing further injury than to spread alarm wherever they were seen.'[16] This sounds like the description of a spectacular auroral display. Coming, as it did, simultaneously with the earthquake, it must have mystified the Jesuits. Perhaps, too, they might not have recognized it as an aurora, since it took place during the Maunder Minimum of sunspots, when little auroral activity seems to have been occurred.[17] The Jesuit fathers may have read of aurora but never had seen an intense display.

As if this were not enough, late in the next year, in 1664, two comets appeared, the brighter one moving north to south and visible from early November until January, the other moving in the opposite direction.[18] An exact account was given by François le Mercier in 1665. At Quebec nine observations were made between 14 December 1664 and 1 January 1665, by which time the comet's tail had virtually disappeared. It was observed until 15 January, but 'the high wind and excessive cold having disturbed our instruments, which we were unable to readjust with all the exactness necessary on such occasions,' their later observations were useless.[19] The second comet, nearly as large, was observed between 29 March and 17 April 1665. No description of the instruments employed is given, but one can imagine that they were simple devices such as quadrants or cross-staffs.

The same *Relation* also calls attention to fireballs seen over Quebec, Tadoussac, and Trois-Rivières that year. According to the *Relation* of 1669–70, an attempt was made to calculate the longitude of Quebec anew. The solar eclipse of 19 April 1670 was observed thus:

It began at a quarter before two o'clock, and ended at twenty-three minutes after three, its total duration being 1 hour and 40 minutes – the whole being measured by the movement of a Pendulum exactly adjusted to the movement of the sun. The extent of the Eclipse was a little more than five fingers. We had marked on a card six concentric circles, separated by equal distances, with each space divided into twelve parts to give us the minutes by fives. But this device being too large for the dimensions of the place where we had taken up our position to make the observation, we were unable to estimate the said extent except by conjecture. If this can serve for determining the Longitude of Quebec, well and good.[20]

It was observed further west by missionaries in the field, but as they lacked instruments and clocks necessary for accurate timing, they could not ascertain the difference in longitude between the mission and Quebec. This was done in many cases with lunar eclipses, but obviously with much less accuracy.

Gaining an advantage over the native population by means of a knowledge of astronomy was not new, for Columbus had done it; so, too, in New France. In the *Relation* of 1673–4 there is an account of the lunar eclipse of 21 January 1674. A missionary challenged the native diviners to predict the time of the eclipse, which they could not do. Armed with an almanac, he could predict the date, time, and extent of the eclipse.. When it occurred as he foretold, 'They were compelled to admit that we knew things better than they ... Such perceptible things have a much greater effect on their rude minds than would all the reasoning that could be brought to bear upon them.'[21] Some of the Jesuits, however, were interested in native culture and recorded some of the main ideas of astronomy of the Indians, preserving them for us in the *Relations*. Father Chrestien le Clercq, who worked among the native people in the Gaspé region, related that they had a lunar calendar, with thirty days in each month, but had five lunations for summer and five for winter, as did the Romans, le Clercq adds, before the Julian reform. Their year began in autumn, but they gave no indication of whether they knew of the equinoxes or solstices, though it seemed to him unlikely.[22]

The Jesuits were interested in all natural phenomena. The Great Lakes fascinated them; in the *Relation* of 1676–7, the writer gives an account of Father André's observations of the tides in Green Bay ('bay des puants'). Father André kept a journal for a year and noted that there were variously two, three, and four tides daily and attempted to deduce the exact connection between the location of the moon and the tidal periods.[23] Astronomy was only one of many activities of the Jesuits, but in a country that lacked scholars they represent the first astronomers in Canada. Education was an early feature of life in New France. No order was better suited for the task, as the Jesuit educational system in Europe was the best available in the early seventeenth century. The ratio studiorum laid down in 1591 prescribed languages, theology, and three years of philosophy, essentially Aquinas and Aristotle, which included mathematics and science. Noteworthy scientific students of the order included Descartes, Réaumur, Buffon, Lalande, and Cassini.[24] When the Jesuit missionaries arrived in Canada it was only natural that, as the population grew, their services as educators would be required. As early as 1635 there arose the Collège de Québec, primarily a secondary school for Canadiens and native children. By 1659, the full course of collège classique was being offered; in the last two years, it included some science and mathematics.[25] Astronomy, if even touched upon in this course of study, would have been only a minor part of the course. Bourdon is said to have taught hydrography in the 1630s,[26] and since at least some astronomy was included in the study of that subject, students at the college may have had an opportunity to hear lectures.

From 1661, Martin Boutet de St-Martin (c. 1616–1683), who had arrived in Canada in 1645, began teaching mathematics for surveying and navigation purposes at the college. The intendant, Jean Talon, requested him to teach pilots in 1666, and Boutet became professor of hydrography in 1671.[27] Rudimentary astronomy was thus available, but probably on a restricted basis. In 1685, J.-B. Franquelin was named professor of hydrography.[28] His noted successor was Louis Jolliet (d. 1700). At the same time, a Jesuit in Montreal was providing courses for pilots in navigation and mathematics: Claude Chauchetière (1645–1709) had erected a gnomon or vertical staff in the garden of the Montreal Seminary to discover the latitude of the town. The professorship of hydrography at Quebec passed into Jesuit hands in 1708[29] and was successively held by nine priests, including Chauchetière; the office terminated with Joseph de Bonnécamps (1707–90), incumbent at the time of the British attack on Quebec.

Bonnécamps seems to have taken more interest in astronomy than most of his predecessors.[30] It was his intention to add an observatory to the roof of the Collège de Québec, similar to those of many Jesuit colleges in

France. On 29 October 1744, Intendant Hocquart wrote the minister of marine on Bonnécamps's behalf to obtain 'une pendule à secondes et une lunette montée sur un quart de cercle.'[31] The cost of an observatory was estimated at between 1,000 and 1,200 francs. Nothing came of the request, and Hocquart's successor, François Bigot, wrote anew in 1748. Bonné-camps now wished to have an additional telescope and a quadrant of 3-foot radius. In the mean time, Bonnécamps transmitted meteorological observations to France. He is best known for his report on the expedition of Céloron de Blainville to the Ohio River in 1749: he was to take observations along the way to fix the geographic locations, but the instruments he had requested from France had not arrived. He remarked that his watch was of poor quality and did not allow him to find longitudes,[32] but he took readings for latitudes in many locations.

The instruments finally arrived in Quebec in 1749, after Bonnécamps had departed with Céloron. In 1752, Bonnécamps travelled to Fort Frontenac (modern Kingston, Ontario) to take astronomical observations. In 1755, with the help of a young officer, Lotbinière, he made new observations to ascertain the latitude and longitude of Quebec. As he reported to J.-N. Délisle, they had found Jean Deshayes's earlier estimate untrustworthy. This was due to his poor instruments, which Bonnécamps now possessed and had examined. They revised Deshayes's value for latitude from 46°55' to 46°48', only 1 minute too small. For longitude, they tried several observations, although they noted clock errors and concluded that Quebec was 4^h 50^m and a bit, or 72°30', west of Paris.[33] This is about 3 degrees too great.

Bonnécamps returned to France in 1759, before the fall of the city, and never returned. Although he was in New France just at the end of the French régime, he can be rightfully called the first important astronomer in Canada. Like his successors of the next century, he fits the pattern of the astronomer in a growing colony – the practical scientist. Although he had the necessary instruments for ten years, we do not know whether his rooftop observatory was ever built in Quebec. Should further research turn up evidence that the instruments, which were clearly portable, were erected, the observatory of the Collège de Québec would have been the first in North America.

Bonnécamps was not the only Jesuit then living in New France interested in astronomy. Pehr Kalm mentions that he met, in 1749, a Father Gillion of Montreal, who had a taste for mathematics and astronomy and had drawn a meridian in the seminary garden 'which he said he had examined repeatedly by the sun and stars, and found it to be very exact.'[34] According to Father Gillion, the latitude of Montreal was 45°27'.

There were no public libraries in New France, nor was there a printing press. Private libraries existed in the homes of the better classes, and the Collège de Québec had a small, but good library. Although the college was closed by the British, its library passed to the Séminaire de Québec and eventually to the Université Laval. Drolet has catalogued the books;[35] of seventy scientific studies that survive, thirteen are astronomical. These range from the earliest, an almanac of John of Seville (1595), to the last, Maupertuis's well-known *Discours sur les différentes figures des astres* (1742). Most of the works are ephemerides or other tables by Argoli, la Hire, Magini, Manfredi, Mezzavacca, and Pagan. Two studies by the early Copernican Phillip van Landsberg are included, and one on the quadrant by Jean Tarde completes the collection. No works of Newton or his contemporaries are included, though it is likely that the Jesuit fathers were aware of him. The books are obviously an adjunct to teaching navigation and for exploration purposes. The professors of hydrography may also have had astronomy books in their private collections. Curiously missing are the standard textbooks of the time, even older, pre-Copernican works such as Sacrobosco's *De sphaera*. Jesuits would have studied such works and may have had them in Quebec, but there are no traces of them. That they were lost is likely, because the books catalogued by Drolet account for only a fraction of Deshayes's bequest of 1706.

One must not overestimate the quality and quantity of education available during the French régime. While certain leaders, especially Bishop Laval, pressed for educational opportunity, there were few schools in the colony. Gilles Hocquart, intendant in the 1730s, reported about 1737:

All the education that most of the children of officers and gentlemen receive is confined to very few things; they scarcely know how to read and write; they are ignorant of the first elements of geography, of history ... At Montreal, the youth are deprived of education altogether; the children go to the public schools which have been established at the Seminary of St. Sulpice and the Monastery of the Charon Brothers, where they only learn the first elements of grammar. Of the young men who have no other aid, they can never become useful men.[36]

As to astronomy, only a handful of students at the Quebec college and the few students tutored by Chauchetière in Montreal would have received even the rudiments. By the time of the Conquest, only 60,000 French lived in the vast area from Quebec to New Orleans, and two centuries of neglect by the mother country had not allowed for a proper system of education. Only later, under British rule, would French-Canadian education blossom and astronomy become available to more than a handful of students.

Until 1760, astronomy practised in the French and in the British dominions of North America differed in intent rather than method. A glance at the map of North America before 1760 shows a continuous ribbon of settlement from Georgia to Newfoundland – with a gap at Cape Breton – in British hands and a band of very sparse French settlement from Louisbourg westward through the Great Lakes and southward to New Orleans. The French possessions were sandwiched by the nominally British territories of the Hudson's Bay Company to the north and west. In what is now Canada, the British population in Newfoundland and mainland Nova Scotia was slight and astronomical activity virtually nonexistent. In the vast Hudson's Bay Company possessions, there was no permanent European settlement apart from trading posts: astronomy practised there was for purposes of exploration and navigation of a Northwest Passage. The fixing of geographic locations and surveying, practised in New France, arose because of settlement and local navigational needs. These needs essentially did not exist in the north.

With few exceptions, the history of astronomy in British North America is the story of latitudes and longitudes. The French had partially mapped their vast American empire, but with the coming of British control and settlement it was to be done again. Nearly all the surveyors and hydrographers of the period, many of whose works have been recounted elsewhere,[37] employed astronomy at some time. Here, we shall deal with a few noteworthy men and their works.

In 1754, the British fleet anchored before Louisbourg in the first major drive to wrest Canada from France. With the fleet were four men who contributed to contemporary astronomy. Samuel Jan Holland (1728–1801) was an engineer with the army and was given the task of surveying the Louisbourg area after its capture. Here he met Capt John Simcoe, RN, father of John Graves Simcoe and master of HMS *Pembroke*, along with Simcoe's sailing master, James Cook (1728–1779). The three became good friends. Simcoe made a number of observations for latitude and longitude along the coasts of Newfoundland and the Gulf of St. Lawrence. According to Holland, Simcoe 'told Capt Cook that as he had mentioned to several of his friends in power the necessity of having surveys of these parts and astronomical observations made as soon as peace was restored, he would recommend him to make himself competent to the business by learning Spherical Trigonometry, with the practical part of Astronomy.'[38] This Cook did, and he aided Simcoe in the task. Simcoe died the next year, but Cook continued his astronomical work in Newfoundland and the gulf during the 1760s, after the fall of Quebec. In 1766, he observed a solar eclipse from Burgeo, Newfoundland, his account being published the following year in the *Philosophical Transactions* of the Royal Society.[39]

Holland remained with the army, becoming an engineer in Wolfe's forces before Quebec. He worked under Gen Murray and in 1764 was named surveyor-general of Lower Canada. He retained this post until his death at Quebec in 1801. Surveying and mapping were his chief concerns, but he made a number of astronomical observations for latitudes and longitudes, including observations of Jupiter's satellites, then a relatively new method for ascertaining longitude. Four of his papers appeared in the *Philosophical Transactions*.[40]

The fourth notable member of the Louisbourg expedition was Joseph F.W. DesBarres (1721–1824), a naturalized Englishman and graduate of the Royal Military College at Woolwich. A remarkable figure, he is best known for his hydrographic survey of Nova Scotia coasts, his maps being published as the *Atlantic Neptune* in the 1780s. He was lieutenant-governor of Cape Breton and governor of Prince Edward Island as an octogenarian. Roy Bishop makes a good case that DesBarres erected the first astronomical observatory in North America at his home, Castle Frederick, near Falmouth, NS.[41] The observatory was built in 1765 for checking surveying instruments. By the 1770s, DesBarres had a reflecting telescope – Bishop believes by Dollond – a refracting telescope, a quadrant, and a Hadley's quadrant. The refractor was a Dollond achromatic with 3.75-inch aperture and a 3.5-foot focus. It is surmised that this telescope was sent from England for the transit of Venus of 3 June 1769. Little else is known of the observatory; it was gone probably by the early nineteenth century. DesBarres published no astronomical papers.

The transit of Venus of 1769 saw one official expedition to British North America. The Admiralty had plans for observing the transit, but not in Canada. The Royal Society sent out four observing teams, one of which went to Fort Churchill on Hudson Bay. The astronomers, William Wales and Joseph Dymond, who was Nevil Maskelyne's assistant at the Greenwich Observatory, took a quadrant, a clock, and two 2-foot reflectors.[42]

While the British were occupying Canada, the vast lands to the north and west of Canada were being explored by men of the Hudson's Bay Company (HBC). Founded in 1670, the company established posts first along the shores of Hudson Bay but moved into the interior of the continent during the next century. Towards the late eighteenth century, it routinely employed men with astronomical training for mapping and exploring, several of them well known to students of early Canadian exploration. In 1769, the company sent Samuel Hearne (1745–1792) westwards from Hudson Bay. Over the next few years he covered parts of the Barren Lands and reached the Coppermine River. His principal astronomical instrument, for finding latitudes, was a Hadley's quadrant made by Scatlif. This instrument, first described by John Hadley in 1732, was the forerunner of the sextant. Its advantage over the astrolabe was that one could 'shoot' the sun without looking directly into the glare. Hadley's first instrument could give readings to one minute of arc.[43] Unfortunately, Hearne does not seem to have employed it well, as his latitudes are generally inaccurate. Having broken that instrument – it fell over in a wind while he was having luncheon – he had to resort to a larger quadrant by Elton which had been at the Prince of Wales fort for some thirty years. He found it cumbersome. Such quadrants, very difficult to transport, were typically used in observatories. As for longitudes, Hearne simply guessed. In 1795 he published an account[44] of his journey, a chatty work of little astronomical importance. His latitude readings were called into question by British hydrographer Alexander Dalrymple, rightly so it seems, and the preface to Hearne's work is an attempted rebuttal.

The HBC had better luck in sending out astronomy teacher Philip Turnor (1751–1800). His trip to Lake Athabaska in 1791 resulted in a slim volume of observations, *Result of Astronomical Observations Made in the Interior Parts of North America* (London, 1794). The book gives latitudes and longitudes for a number of positions and includes observations of his pupils, Peter Fidler and David Thompson, along with observations by Alexander Mackenzie and Capt James Cook. Fidler (1769–1822) succeeded Turnor as the company's chief surveyor on the latter's return to England. Alexander Mackenzie (1764–1820), whose exploring activities are well known, was in London when Turnor was instructing the others, himself learning more astronomy and surveying techniques. Armed with a new sextant, chronometer, and telescope, he made his trip overland to the Pacific in 1792–3. He made a number of latitude observations but took care to record the longitude near the ocean; this was done by timing Jovian satellite eclipses. His account was printed in 1801.[45]

The most remarkable and painstaking of the HBC astronomers was David Thompson (1770–1857). He studied with Turnor in 1789–90 and made a series of exploratory trips until he left the HBC for the North West Company (NWC) in 1797. For that company he observed latitudes and longitudes of posts and surveyed the 49th parallel in the Lake Superior area.

In 1815 he retired from the NWC, and from 1816 to 1826 he was a member of the International Boundary Commission, established by the Treaty of Ghent in 1814.

For the exploration of the Pacific coast, two men should be noted.[46] Capt Cook touched at Vancouver Island in 1778 on his last voyage. He found the position of Nootka Sound from a temporary observatory. It was to this place in 1792 that Capt George Vancouver (1757–1798), a midshipman under Cook and a student of the astronomer William Wales, took an expedition to survey the coasts of the Northwest. The astronomer sent out with the expedition, one Mr Gooch, having been killed in the South Pacific, Vancouver had to perform the tasks himself and had twelve sextants by Troughton, Dollond, Ramsden, and others, along with chronometers. Establishing a temporary observatory on Vancouver Island, he was able to obtain the longitude by lunar observations.

Mention should be made also of the early boundary surveys made in British North America. While much of the work was ordinary surveying practice, astronomical observations for latitude and longitude were necessary. The boundary line between the United States and British North America was in dispute for nearly half a century from the end of the American Revolution. After the signing of the Treaty of Ghent ending the war of 1812, the British sent out Dr J.C.C. Tiarks with instruments supplied by John Pond, the astronomer royal.[47] This was insufficient, and in 1843, as a result of the Webster-Ashburton Treaty, American and British teams surveyed the boundary from the Maine–New Brunswick line to Quebec. The British team was led by Lt-Col J.B.B. Estcourt (1802–1855) and was equipped with a full set of instrumentation. Astronomer Royal George Airy had instructed the team to take a number of chronometers, a method for land longitude measurement that he had already initiated in Britain and would carry on in co-operation with the Harvard College Observatory. Estcourt's team obtained the longitude of Quebec Citadel by lunar distances. Estcourt's values, as compared with those of the Americans, would create controversy later, when newer techniques superseded the old.

As the surveying of geographical positions continued apace, work began on the Great Lakes and Atlantic coasts. The hydrographic surveys of brothers Sir Edward Owen and Capt William Owen, commencing in 1815, and of Lieut (later Admiral) Henry Wolsey Bayfield (1795–1885) employed astronomy regularly.[48] Bayfield worked on the Great Lakes from 1816 to 1825, then returned to Lower Canada in 1827 to survey the St Lawrence. He found the longitude of Quebec by observing occultations and eclipses of

Jupiter's satellites. Using Quebec as his secondary meridian, he employed chronometers and sextant for other longitude points. His value for the longitude of Quebec was the accepted value until Estcourt's determination.

The advent of British rule over New France spelled an end to the Jesuit system of education. The Collège de Québec had closed during the bombardment of the city in 1759 and was then requisitioned by the British army as a barrack. The expulsion of the Jesuit order from Canada in 1763 meant the end of the institution: Bonnécamps's earlier departure had effectively ended any teaching of hydrography and astronomy. French-Canadian schools soon reopened, but astronomy disappeared from secondary or technical education until the late 1790s. There is no evidence that the poorly educated parish priests of the post-Conquest era carried on the Jesuits' interest in astronomical phenomena.

The significant influx of English-speaking settlers did not commence until 1783, at the conclusion of the American Revolutionary War. The Loyalists, many of whom were educated, had little time in their early years in Canada for higher education or scientific pursuits. The next wave of settlement, at the turn of the century, included American farmers and retired British soldiers, who were even less inclined to pursue astronomy for either practical or intellectual ends. The Protestant clergy, with few exceptions, shared little of the intellectual tastes of the Jesuits. Thus, early post-Conquest astronomy, for purposes of land, boundary, and hydrographic surveying, was in the hands of British-trained men such as Holland, Bayfield, DesBarres, and Estcourt. Land surveyors came from either Britain or the United States or were trained by such men locally in an apprenticeship-style system. There were no schools of surveying, no colleges offering scientific instruction, and no government-supported technical courses such as the French chair of hydrography. The formal French system, slight as it was, gave way to the informal British approach.

The colonial era in Canada can be divided into two periods, before about 1840, and after, till Confederation. In the earlier period, government was in the hands of officials appointed by the home governments of France and Britain. When astronomy was practised in an official capacity, it was done so at the behest of imperial governments, usually for their own needs and by their own men. The colonists themselves rarely initiated or performed any astronomical work. Only a handful of colonists ever learned the rudiments of hydrography, surveying, or astronomy. After 1840, as responsible government was introduced and colonial legislatures assumed more control over their affairs, astronomy practised for locally perceived needs began to make its appearance. At the same time, with the slow improvement of Canadian education, Canadian-born and -educated men could begin to cultivate astronomy.

The scientific work of the Hudson's Bay Company and the British arctic explorers, and official expeditions like those of Chabert, Wales and Dymond, Cooke, Vancouver, DesBarres, and Bayfield, were examples of astronomy in Canada. All such work was done in accordance with needs perceived beyond Canadian shores. To a great extent, the valuable reports of the Jesuit missionaries were a response to the philosophy of their order rather than a systematic effort to create an inventory of the natural environment of Canada to assist colonization and settlement. Thus astronomy, as one small scientific aspect of nation-building, belongs to the period after 1840, when we can begin to think about Canadian astronomy. The state would play a central role in that story, and the state would be increasingly the Canadian state rather than the imperial. This movement will be detailed in the next chapter.

PART TWO
Canadian Astronomy
1840–1905

William Brydone Jack, president, University of New Brunswick;
from a portrait at the university

Members of International Boundary Commission, 1893–5;
several later joined the Dominion Observatory.

Canadian eclipse expedition, Jefferson, Iowa, 1869;
Cmdr Ashe at eyepiece of Quebec Observatory refractor

Capt Charles Darrah, RE, British boundary survey team,
with zenith telescope, Yalik River, 1860–1

Otto J. Klotz, early 1890s

Toronto-area amateurs, c. 1900

Quebec Observatory and time ball, 1870s

McGill College Observatory, early twentieth century

Second Toronto Observatory, constructed 1857, with added dome
for Cooke refractor, c. 1905

CHAPTER TWO

Government and Astronomy

The evolution of astronomy in Canada parallels the economic and social growth of the country. Before about 1840, we must speak of astronomy in Canada; after that point, we can begin to perceive, dimly at first, a uniquely Canadian approach to astronomy, which, during the second half of the century, becomes more and more apparent. So it was with Canadian nationhood. Before the 1840s, British North America was a loose collection of colonies with limited local government. Legislative assemblies had few financial resources and little power. Authority lay with the local lieutenant-governor, select officials and families that supported him, the governor-general, the army, and the Colonial Office. Because science costs money and local initiative could be only slight, scentific activity was almost entirely due to imperial initiative.

Nor can we speak, in this period, of 'Canadian' needs, for the separate colonies, distinctive culturally, were at varying levels of development. In Lower Canada, the francophone majority had a culture with deep roots, while in Upper Canada, the past encompassed only a half-century. The years surrounding 1840 were critical to the formation of the Canadian nation: as one result of the report of Lord Durham, Upper and Lower Canda were arbitrarily joined by an act of 1840 to form the Province of Canada, the nucleus of Canada as we know it. With a larger population and land area and greater economic strength, central Canada exercised an attraction – sometimes vigorously opposed – on the smaller Maritime provinces, leading to Confederation in 1867. Central Canada was also the staging ground for the exploration and population of the vast western regions. Most scientific activity occurred in central Canada, and the evolution of Canadian astronomy is focused there.

The year 1840 also marks the beginning of the move to responsible government in Canada. Colonial assemblies were now expected to exercise more authority and, naturally, to assume a greater financial burden, a result probably of imperial policy more than local desire. Over two decades, the imperial government increasingly withdrew from Canadian affairs, with consequences for science. The Toronto Observatory, established by the British in 1840, was abandoned to the province by 1853; the Geological Survey of Canada, the world's second oldest national survey, was created in 1842 with no imperial assistance at all. The government in London had a hand in the formation of several other observatories or surveys, but the local governments were expected to maintain such operations with little or no imperial contribution. As a result, except for the 49th-parallel boundary survey, virtually all astronomical initiatives in British North America after 1840 were in the hands of the colonists.

A striking feature of the growth of Canadian society and its economy, and equally of the sciences, is the significant role of the state. For several reasons, public enterprise was essential to the evolution of Canada from at least the 1820s.[1] Security could be provided only by the state, given both real and perceived danger from the United States. The enormous geographical expanse of Canada, compounded by the fact that the habitable portion was a thin band stretching along an east-west axis, punctuated by difficult terrain, meant that transportation and communications would always be challenges. To those difficulties one can add another, a small, dispersed population. Canada numbered only some three million at Confederation, and only the state could afford to support roads, canals, railways, and, ultimately, industry. In contrast, the growth of American society and its economy, even with its family resemblance to Canada's, relied far more upon individual enterprise. Of course, the state also intervened in the United States, but the need was far less, given several important advantages: far more habitable area, more easily penetrated terrain, a far greater population, and a longer history. Naturally, industrialization evolved much faster there, and private fortunes, through philanthropy, could be devoted to higher education and to science. Astronomy was a particularly important area for philanthropy in the nineteenth-century United States.[2] Many Americans have believed in the importance of private enterprise and philanthropy; proportionately fewer Canadians have, and when several Canadian industrialists did amass fortunes, only a handful funded the sciences. It seems that rich Canadians believed in the role of the state as much as did their poorer countrymen. Scores of US observatories and laboratories were created with gifts during the last century, while the first Canadian example – the David Dunlap Observatory – had to wait until the 1930s.

Also, university education was developed more widely in the United

States, in terms of both number and size of institutions. In the sciences, during the nineteenth century no Canadian university could be compared with Harvard, Yale, or the Massachusetts Institute of Technology. Thanks to philanthropy and to larger institutions, American astronomy was centred on universities to a degree never matched in Canada. State-supported astronomy, then, was the core of nineteenth-century Canadian astronomy; the science, as part of university education and of general culture, was far less developed in Canada. That aspect will be deferred to the next chapter.

In a small and developing country, practical science is of far greater value than scientific knowledge for its own sake. Where astronomy could benefit Canadians, it was supported by governments, increasingly so as needs grew. Government-supported astronomy in nineteenth-century Canada developed through three main phases. In the first phase, colonies provided small grants to individuals or groups for specific purposes, such as surveys or the determination of longitudes and latitudes for surveying purposes. Before the 1860s, none of these grants was permanent or amounted to much in monetary terms. The legislators in Canada or in New Brunswick, for example, expected a reasonably quick return on their investment; they did not intend to create long-term astronomical programs. Astronomy was not alone. The Geological Survey of Canada was meant to be a one-shot affair, to provide a quick overview of Canadian mineral potential for investors and mining operators. Yet, the complexity and sheer extent of Canadian geology, combined with the political acumen of Sir William Logan and his successors and supporters, not only maintained the survey but facilitated that non-economic activity, pure research.

In the second phase, after Confederation, a more permanent approach emerged in government astronomy, when the small observatories financed formerly by the Canadian legislature were drawn together into the Canadian Meteorological Service. Here, again, practical needs were paramount: accurate time for cities and railways, and weather information for agriculture and shipping.

The third phase, which would lead to a permanent branch of government devoted exclusively to astronomy, arose out of the need to survey the newly opened Canadian west: the Astronomical Branch of the Department of the Interior appeared in embryonic form in the 1870s.

This chapter will detail the evolution of state-supported astronomy, with special attention to the Quebec Observatory, which, under the direction of E.D. Ashe, was the only nineteenth-century institution to include 'pure' research alongside the eminently practical activities of its sister insitutions.

During the eighteenth century, the practice of astronomy in Canada was almost exclusively directed towards geographical needs; these requirements remained essentially unchanged for the first half of the nineteenth century. British North America was a collection of small provinces with separate governments and different cultures. There was little co-ordination of efforts except by whim of the Colonial Office in Whitehall. Most of the Canadian landmass was in the private corporate hands of the Hudson's Bay Company, and populations were small and unevenly distributed. The cultivation of astronomy as either professional or amateur activity required a population large enough to sustain interest, sources of funds, and enough trained or educated individuals, but these factors were generally missing in the first half of the century. None of the largest towns – Halifax, Quebec, Montreal, Kingston, and Toronto – exceeded 30,000 in 1830. Therefore, astronomical activities before about 1860 were relatively isolated, men working alone, having little contact with one another or with astronomers in the United States or in Europe. The government's interest in astronomy was limited to modest geographical or geophysical surveys. The pattern of provincial government astronomy in the first half of the century was carried on in the latter half by the federal government.

The first example of local government astronomy in Canada is one of failure: the attempt to establish an observatory in Toronto with the instruments from St Helena's observatory, which had been dismantled about 1835. In the account of this episode by Douglas,[3] it is evident that the idea of an astronomical observatory in Canada was premature before the 1850s. The notion seems to have been that of John Harris, a resident of the London district, who in 1833 petitioned Upper Canada's assembly to erect an observatory in York (Toronto). The Admiralty agreed that an observatory was an extremely worthwhile project, there being none in North America nor any standard meridian from which to base longitude work. The St Helena instruments could be provided, including two transit instruments, four telescopes, a variety of surveying instruments, four clocks and chronometers, and meteorological instruments. Lord Glenelg, the colonial secretary, wrote Sir John Colborne, lieutenant-governor of Upper Canada, on 29 July 1833 urging that the project be undertaken. Colborne requested a report on a location from Capt Richard Bonnycastle, RE, who submitted the document on 10 November and made a strong plea for Toronto as the site. Colborne, in a letter to Glenelg of 23 November, seemed to assume that the Admiralty would pay the salary of a professor and urged Glenelg to have such a person sent out.

The ensuing exchange reflects a permanent feature of British-Canadian relations in the early nineteenth century. The Admiralty passed the problem to the Colonial Office, which passed it on to Colborne, asking whether the legislative assembly would defray the costs of an observatory. Nothing happened. Glenelg asked Colborne's successor, Sir Francis Bond Head, to take the matter up with the legislature, but he seems not to have done so. On 21 May 1837, the Admiralty reported to the Colonial Office that Airy, the astronomer royal, had seen to the repair and packing of the instruments and that they were ready to be sent. Still nothing happened. While it would be easy to blame Head for his inaction, the brewing troubles, which would culminate in the Rebellion of 1837, were occupying the minds of those in the Upper Canadian government and legislature; an observatory was certainly the furthest thing from the lieutenant-governor's mind. When the political situation settled down, a new proposal superseded the earlier one, with much wider sympathy and backing.

The science occupying minds in Britain was geophysics, specifically terrestrial magnetism. In 1839, after memorials from the Royal Society and the British Association, the authorities decided to make a concerted effort to obtain magnetic observations in a systematic fashion, following the lead of some German and Italian states, France, and Russia. Under the overall guidance of Maj Edward Sabine, imperial observatories were established at St Helena, Capetown, Hobart (Tasmania), and Toronto. The Toronto Magnetic Observatory, built in 1840 on land provided by King's College, was under the direction of Lieut C. Riddell, RA.[4] The observatory was not intended for astronomical work, although it had a transit hut attached to the main building. A transit theodolite was the sole astronomical instrument. In 1842, Lieut John H. Lefroy, RA (1817–1890), was appointed director. As one of his first acts, he decided to undertake a magnetic survey of the Northwest.[5] Departing Montreal in May 1843, he reached Fort Chippewyan, where he wintered, and then proceeded to Fort Simpson on the Mackenzie in the spring. With one assistant, he made a great number of observations, publishing them soon after his return to Toronto.

In 1853, the imperial government decided to cease operations at Toronto as part of its general withdrawal from Canadian affairs. The youthful Canadian Institute petitioned the provincial government to assume responsibility for the observatory so that it could continue its magnetic and meteorological work but also noted 'that the time has rather come when its operations should be placed upon a less restricted basis, and to be extended ... to include those higher departments of science, and more particularly of Astronomy, to which every Canadian must aspire to see his country one day contribute.'[6]

Lefroy interceded with the governor-general, and the legislature agreed to finance the observatory. Local government sponsorship of the institution was predicated on practical needs – meteorology – and growing feeling that support of science reflected the country's maturity. John B. Cherriman

(1823–1908), professor of mathematics and natural philosophy at the university, assumed the directorship, and the work went on as before. In 1855, George T. Kingston (1817–1886) was named professor of meteorology at the university and succeeded Cherriman at the observatory. Astronomy, apart from the provision of time locally, did not figure in his work.

The birth of what would be called professional astronomy in Canada occurred in Nova Scotia and New Brunswick in the 1850s. Because those colonies were separately governed and culturally different from the two Canadas, there was little direct connection between events in the Maritimes and the slow coalescence of government astronomy further west. The east had few urban centres, few railways, and little industry or agriculture on the scale of Canada. French-English tensions were lacking, and in Nova Scotia the English population had been in place longer than that in Canada and had strong similarities to its New England counterpart.

Capt Bayfield had already made many astronomical observations in connection with his hydrographic survey of the St Lawrence. By the 1850s, he and Cmdr P.F. Shortland were mapping the Bay of Fundy and the Gulf of St Lawrence and required a reference meridian.[7] This was obtained from Harvard College Observatory by means of telegraphy. The method of exchanging signals of star transit observations by telegraph had been pioneered in North America by William Cranch Bond, director of the Harvard College Observatory, though the method had been employed earlier in England. In 1852, exchanges of signals between Halifax and Cambridge, Mass, via Bangor, Maine, led to establishment of Halifax's longitude. The longitude of Harvard Observatory had been earlier ascertained by means of the exchange of chronometers with Greenwich, since the trans-Atlantic cable, though projected, had not yet been built.

In Fredericton, William Brydone Jack (1819–1886), professor of mathematics and natural philosophy at King's College, learned of the Halifax results and wished to improve upon his own longitude work. He had already made the acquaintance of William Bond in Cambridge. In a letter to Bond in July 1854 Jack mentioned that he had calculated the longitude of Fredericton by means of lunar culminations.[8] By this method, the position of the moon's limb was determined from nearby stars, listed for that day in the *Nautical Almanac*. The difference in time between two places could then be calculated. The two men agreed to collaborate, employing the new technique.

Observations were made during January and February 1855 at the small observatory of Jack's friend, Dr James Toldervy, a Fredericton physician. The observatory, located near the embankment of the Saint John River and close to the telegraph station, was used in preference to Jack's new college obervatory on the hill, which was not close to a telegraph line. The longitude determined was 66°38′21.5″w, about 18 seconds different from Jack's earlier calculation from lunar culminations. Later that spring Jack began work with a transit instrument and chronometer lent him by the Canada–New Brunswick Boundary Survey Commission.[9] During 1855 and 1856, Jack and Toldervy observed from a number of sites in New Brunswick and were able to fix the longitudes by the telegraphic method.

Writing to Sir George Airy in October 1856, Jack relates his and Toldervy's attempts to improve the efficiency of their work. He notes that Toldervy's transit, a 30-inch Simms, had had an extra prism added but that, since it had only a glass reticle with scratched lines for cross-hairs, it would require proper spider wires. They would have to effect this themselves because there were no instrument-makers in New Brunswick. Jack describes his situation – one shared by most nineteenth-century Canadian astronomers – thus: 'I must beg pardon for troubling you with these crude remarks, which I trust your kindness will excuse, seeing that in this corner of the world I have seldom a chance of holding intercourse with Astronomers of repute. I wish I had more of such intercourse so that I might profit by a constant interchange of ideas and be stimulated and enabled to do something to make me regarded as a not unworthy votary of astronomical science.'[10] Being isolated as he was, he found himself in a quandary when he could not get his results to match those of the earlier boundary survey, performed in compliance with the Ashburton-Webster Treaty of 1842.

Jack's observations at Grand Falls (now Edmundston) gave him the first intimation of the problem. Connected telegraphically with Toldervy in Fredericton, Jack sent signals at 1-second intervals from his chronometer. These were recorded in Fredericton, and a set of signals was sent in the other direction. After two exchanges either way, the men were able to find the mean difference in sidereal time between the two places. With their transit instruments, they attempted to find the clock errors. When these were compared, the longitude of Grand Falls with respect to Fredericton, and therefore with respect first to Harvard, then Greenwich, was known.[11] Jack could do nothing about their personal equations. He did find that the longitude of Grand Falls as recorded by the American boundary survey team under Maj Graham was extremely close to his own, while the British value, ascertained by Gen Estcourt's team, was nearly two miles different. Airy was annoyed: 'The ensuing accord of your determinations with those of the American authorities, may tend to raise disagreeable feelings in the minds of some of the persons concerned.'[12] It was his belief that one had to know which method was more accurate: Jack's and the Americans' use of Harvard chronometric time, or the British observations of lunar transits. Jack had to reply that the British determinations were admirably done, but Toldervy, from lunar transit observations in Fredericton in 1846–7, had found a value about 10 seconds less than he and Jack subsequently calculated with Bond's aid.[13] The subject was closed when Airy advised the Admiralty not to publish Jack's memoir.[14]

Jack had had no one in Canada with whom to collaborate. With the arrival of E.D. Ashe at Quebec, a connection between the Maritimes and Canada was possible. However, in the Northwest, where there were few settlements and no telegraph lines, longitudes had to be measured with the use of transit observations and chronometers. Two expeditions from Canada into the Northwest took place in the 1850s. Capt John Palliser led a British expedition to the prairies in 1857, taking with him J.W. Sullivan as astronomical assistant. This team went on to the Rockies in 1859–60. At the same time, a provincial expedition under Henry Youle Hind (1823–1908), a professor at Trinity College, Toronto, explored what is now Manitoba and Saskatchewan. Of more astronomical interest was the survey of the 49th parallel, undertaken by British and American teams,[15] who had successfully surveyed this line west of the Rockies from 1857 to 1862. The remaining portion of the line, between the Rockies and the Lake of the Woods, was surveyed in 1872–6. In the latter effort, Lieut Samuel Anderson, RE, was appointed chief astronomer for the British team. Canadians were appointed as well, including William F. King (1854–1916), later Canada's first chief astronomer; Dr George Mercer Dawson, later director of the Geological Survey; and W.A. Ashe, later director of the Quebec Observatory. King and Ashe were assistant astronomers for the team. Locating the boundary was essentially a problem of determining latitudes and surveying, but longitudes were also wanted. Since the telegraph had recently been extended to Pembina, North Dakota, clock signals were exchanged with the Chicago Observatory and the needed longitudes were determined. Once again, for lack of a national observatory, Canadians had to rely on the Americans.

Before 1850, the only government observatory in Canada was the Toronto Magnetic and Meteorological Observatory, which was not, properly speaking, an astronomical observatory. Astronomical surveys carried out for geographical purposes by Bayfield and others were modestly underwritten with government monies, but astronomy was not a priority to either imperial or provincial governments. This situation changed in 1850 with the establishment of the Quebec Observatory, the first of a small network eventually drawn together as the Canadian Meteorological Service.

Attempts to obtain an observatory for Quebec can be traced back to at least 1844, at the time of Bayfield's work in the city. Capt Boxer, the harbour master, along with Sir Richard Jackson, commander of the forces in Canada, and the board of trade of Quebec, recommended that an observatory was essential to provide time to shipping. Airy at Greenwich concurred. Correct chronometer time was available in British ports such as Liverpool, but no means existed to provide time at Quebec, then the chief Canadian port. Since longitude at sea required accurate chronometric time, the establishment of an observatory would be a great boon to sailors. Therefore, Earl Grey, the colonial secretary, wrote Lord Elgin, the governor-general of Canada, on 26 March 1847, saying that the plans and estimates were ready to draw up but that 'I incline at present to think, that the cost of erecting the proposed building, together with the charge of maintaining it, should be defrayed by the Legislature of Canada.'[16] The estimate of £526 15s 5d sterling and a materials list were issued in November 1847. The building, to be built in stone with a ball tower, was to be located in Mann's Bastion of the citadel. Airy wrote that the Greenwich Observatory would supply a transit instrument and a telescope of 42-inch focal length, and that the Canadian government ought to purchase at least one good clock but that a second would be desirable.[17] The Canadian executive council was, by 1849, willing to undertake the project, provided that the instruments would be supplied by Britain.[18] The Royal Engineers were given the task of construction, and Earl Grey was advised in May 1850 that the instruments could be shipped and that a half-pay lieutenant of the Royal Navy should be dispatched to take charge of the observatory.[19]

The astronomer chosen, Edward David Ashe (1813–1895), had served in the Royal Navy since 1830 in the Mediterranean and the Pacific but had been invalided back to England.[20] On 18 November 1850 Ashe arrived in Quebec.[21] Airy had meanwhile collected the instruments, a 30-inch transit, the 42-inch telescope, a sextant, and two clocks, one by Dent and the other by Molyneux. These arrived with Ashe, but the time-ball machinery was still in transit. The building had already been completed, and Ashe took charge of it on 27 November 1850.[22] During the next year Ashe retained an assistant for mechanical work and placed the clocks and instruments in their places, but the ball machinery had yet to arrive. Following several changes to the building, including the addition of a revolving turret, the ball machinery arrived and was erected in July 1852.[23] By this time the province, not the imperial government as sometimes stated, had provided more than £2,100 for completion of the observatory; thereafter, annual grants were held to £400 for salaries and supplies, rising to $2,400 by 1857 and remaining at that level till Confederation.[24] This was a much larger amount than the grants to other subsequent observatories. Such government largesse for astronomy would not be repeated until the building of the Dominion Observatory.

Ashe was asked to undertake a number of commissions for the government in addition to operating the time-ball. His own work began in earnest in 1855. On hearing of Jack's work in New Brunswick, Ashe contacted him and arranged for the exchange of telegraph signals to help establish the longitude of Quebec. While a longitude value had already been fixed by Bayfield in the 1840s, it was deemed insufficiently accurate. Ashe had the Royal Engineers bring the lines of the British-American Telegraph Company to the observatory, and he first connected with Jack on 15 November 1855. After one exchange in either direction, Ashe deduced the longitude for Quebec to be $4^h44^m48.59^sw$.[25] While neither Jack nor Ashe felt this to be satisfactory, Jack was pleased to see that their determination was closer to the American value for Quebec than had been the British commission's determination.[26]

At this point, Sir William Logan, director of the Geological Survey of Canada, entered the picture. A clause in the act creating the survey assigned it the duty of obtaining the latitudes and longitudes of important places in Canada,[27] but his small staff had neither the time nor expertise to undertake such work. Statutory limitations did not allow Logan to explore the geology of the neighbouring provinces of British North America, but his desire for broader knowledge had led him to rely on local geologists in New Brunswick, Nova Scotia, and Newfoundland. This same method would work for astronomy. Ashe was approached to find the positions of several points in Canada and asked to co-operate with Bayfield in Halifax and with Jack in Fredericton. To clear the way, Logan wrote W.C. Bond at Harvard to obtain his assistance, which was forthcoming.[28]

Before the exchange of signals between Quebec and Cambridge was effected, Ashe set about connecting Quebec with Logan's Canadian points.[29] In December 1856 and January 1857, he exchanged signals with Quebec from the Toronto Observatory; in February, he worked at Kingston with Dr Horatio Yates; in March, he was in Montreal; and in April, he was in Chicago, working with Lt-Col Graham, of the US Army. After his return from Chicago, he wrote Bond asking for instructions.[30] Bond replied that they ought to exchange clock beats rather than signals corresponding to star transits, since only he had an automatic recording device.[31] Some signals were exchanged in June and July 1857, but Ashe had to go to Collingwood and Windsor to connect those places with Quebec. The final connections between Cambridge and Quebec were made in September and

October. The Quebec Observatory's longitude was found to be $4^h44^m49.0^s$, thus 0.4^s greater than the Ashe-Jack determination.[32] Jack, in Fredericton, was satisfied with the results. His own dispute with Airy on the basis for British North American longitudes was cleared up in the same year, when Bond sent Jack a summary of the determinations of Harvard's longitude.[33] These were transmitted to Airy, who wrote the Admiralty: 'The United States' result ought to be received without any qualification for the British provinces. First, it cannot be very erroneous. Secondly, whether it be right or wrong, it is far more important that continuous districts (British and u.s.) of North America should agree in their basis of Longitude than that the British Provinces should agree with Greenwich.'[34]

Once the longitude effort was finished, Ashe settled into seeing that the time-ball was dropped each day, but he was a far more resourceful and intellectual man than such routine work required. This was already evident in his gunnery researches as a young naval officer.[35] As early as 1855 he began lobbying for a better observatory in Quebec. In his annual report of that year, he comments: 'By mounting an Equatorial, the establishment would be turned into a first class Observatory. This appears to be very desirable, when it is remembered that there is no Public Astronomical Observatory in Canada ... whilst most other countries are contributing to the advancement of Astronomy.'[36] In November 1856 Ashe wrote the provincial secretary suggesting a new observatory. In April 1857, Gov-Gen Sir Edmund Head submitted to the legislative council a proposal for such an observatory, along with another for an observatory for the University of Toronto.[37] Head had earlier been lieutenant-governor of New Brunswick and was a good friend of Jack and a staunch supporter of science. Nothing came of the proposal, but by the next year Ashe was still sanguine. He wrote to his friend Thomas Roy at the Université Laval, thanking him for his help in the project and noting Head's advocacy, as well as that of the Canadian Institute in Toronto. However, 'The opposition to my plan is made by [John A.] Macdonnald (sic), who does not understand, and therefore does not care much about science, but it would be a thousand pities if Canada should be prevented from keeping pace with the enlightened world, through the ignorance of one individual.'[38] Ashe's attempts to branch out finally met with some success. In May 1860, George P. Bond, the new director of the Harvard College Observatory, wrote him to say that the Smithsonian Institution was planning an expedition to Cape Chidley on the northern tip of Labrador to observe the total eclipse of 18 July.[39] Ashe was able to obtain funds to participate. He was the only Canadian member of the team led by Alexander D. Bache of the us Coast and

Geodetic Survey. Taking the Quebec Dollond refractor and transit instrument with him, he was able to secure nearly complete observations of the eclipse.[40]

Although a few Canadian amateurs may have viewed the eclipse, the only official inland expedition was an American group of three, including Prof W. Ferrel and the Nova Scotian-born Simon Newcomb.[41] Working from H.Y. Hind's report of the Assiniboine and Saskatchewan exploration, they found a suitable observing site on the Saskatchewan River in Manitoba. Ferrel's account was published the next year in the *American Journal of Science*. Renewed effort was made in 1861 by Cmdr Ashe to obtain a new observatory for Quebec. Finally, in 1865, he was given permission to order an equatorial.[42] This he did from Alvan Clark, who provided one of 8-inch aperture and 9-foot focal length for $2,500. A wooden building with dome was erected in Bonner Pasture, just east of the old Quebec jail, in the same year. Not until May 1874 was the entire observatory moved to the Bonner site.[43] Once it was in place, this new telescope became the largest in Canada; there were few of larger aperture in the United States.

With such an instrument, Ashe could have turned to one of the important fields of that period – observing double stars or planets, seeking comets and asteroids, or solar work. The last was just coming into prominence with Warren de la Rue's research at the Kew Observatory. Ashe had read with great interest the articles on solar physics by Richard Carrington and others in the *Monthly Notices* of the Royal Astronomical Society (RAS). Prior to 1864, he had been observing sunspots with the Dollond refractors, and once the Clark telescope was mounted, he made them his special study, relating his thoughts on them in several articles in the *Transactions* of the Literary and Historical Society of Quebec, of which he was president several times. In the mid-1860s, the consensus was that the spots were depressions in the photosphere, possibly akin to hurricanes. Ashe's observations led him to doubt this. He had observed the sunspots' umbras forming before the penumbras and had seen umbras fragment and then fade. From this he concluded that the spots were caused by the infall of meteoritic material or small asteroidal bodies. The umbra represented the molten metals and the penumbra the dross that spread outwards. The sun was nebulous, he agreed, so that the granulation of the photosphere was caused by the infall of material.[44] Ashe's theory, then, revived the older view that sunspots are above the photospheric layer.

He reiterated his argument in 'The Physical Constitution of the Sun,' published in 1869, saying that there was a zone of asteroidal bodies between the sun and Mercury which accounted for the perturbations claimed by Leverrier in his search for the supposed planet Vulcan. Ashe argues that, at the sunspot maximum just before 1869, the inner planets were ranged to one side of the sun. This alignment caused greater perturbations in this swarm of material and, thereby, hastened the infall.[45]

During this time, Ashe obtained equipment to undertake celestial photography, which had never before been attempted in Canada. He reported to the *Monthly Notices* in January 1869 that he had mounted a Voigtländer camera on the equatorial, giving him a 3.4-inch–diameter image of the sun at 14-foot focal length. His photographs were good enough to show granulation.[46] Because Ashe was in at the beginning of this new field, there was no one in Canada to assist him. He therefore wrote to Hermann Vogel, editor of the *Photographische Mittheilungen*, in Berlin, from whom he received expert advice.[47]

The total solar eclipse of 7 August 1869 provided an opportunity to photograph the corona and possibly the much-discussed prominences. This was still a highly experimental undertaking; De La Rue had succeeded in getting good wet-plate photographs of an eclipse less than a decade earlier. Ashe was able, after much delay, to obtain $400 in government funds to observe the eclipse. The site selected was Jefferson, Iowa. Joining him in the Canadian Eclipse Party were James Douglas jr (1837–1918), prominent Quebec chemist, mining consultant, and student of the new art of astronomical spectroscopy, and an Englishman, Alexander Falconer; joining them in the United States was a Mr Hugh Vail of Philadelphia. Ashe took the 8-inch and the Dollond refractors. At the site, he took charge of the 8-inch refractor, slightly damaged in transit, while Vail observed with one of the Dollonds. Douglas worked in the dark-room, and Falconer acted as the 'shutter' by holding a cloth over the objective lens. Four plates were made, and the resulting reports led to controversy and much bad feeling.[48]

Ashe claimed to have seen a prominence shoot out rapidly, then bend over before falling back; one plate seemed to indicate this. Vail agreed, as did other naked-eye observers at the site. None of the nearby American expeditions saw or photographed it. The plates were sent to De La Rue, who responded that the telescope had obviously moved. Ashe had also claimed to have photographed concentric luminous envelopes about the sun's limb and notches beneath the prominences. Those were dismissed by De La Rue in the *Monthly Notices*: 'There is evidence of the disturbance of the Telescope during the exposure of the sensitive plates.'[49] As the best eclipse of 1870 was not visible in North America, Ashe requested $800 from the minister of marine and fisheries to take the telescope to Gibraltar, but Peter Mitchell was 'not aware that the interests of Canada require an expedition of this description.'[50]

Not being allowed to follow up his work, Ashe wrote to the Scottish astronomer Lord Lindsay, who had photographed the eclipse from Cadiz, asking whether he had seen any of the phenomena. Lindsay replied that he had seen no stratification of the prominences but had recorded concentric bands, notches beneath the prominences, and at least one blown-over prominence.[51] After seeing Ashe's photographs, Lindsay remarked that they were better than anyone's he had seen, save those of Rutherfurd, and 'it quite beats anything of Kew Observatory I have ever seen.'[52] Lewis M. Rutherfurd, a New York amateur, was an important pioneer in astronomical photography and spectroscopy, using a 13-inch refractor. Ashe says in his observatory report from 1871 that he hoped to surpass Rutherfurd if funds were made available.

Late in 1871, the council of the Royal Society passed a resolution 'fully recognizing the general value of the work in which Captain Ashe is engaged.'[53] Kew Observatory had asked him to co-operate in sunspot studies, and Ashe had made a series of observations from 1868. None the less, the cold reception in England of his theory of the sun had altogether soured him. As he intimated to the Toronto amateur Andrew Elvins in 1871,

I am only too glad to see anyone trying to solve [your] difficult question, and everyone who fairly gives his ideas on any subject without 'trimming his sails' to the opinion of *big men* will do science a service. I read a paper on 'Solar Spots' at the Astronomical Society when in London, and De La Rue was in the chair, and because my opinions were quite opposed to his he said that 'Commander Ashe' was not up in the literature of the sun. That was five years ago, but now I can tell him that these five years' close study have confirmed my ideas, whist they are obliged to *stagger* and get into all sorts of difficulties and contradictions to account for the several phenomena, whilst our views explain everything.[54]

The leaders of British science discounted colonial scientists, but Ashe's abrasive personality may explain De La Rue's dismissal of him. Money was a problem, however, and enough for photography was never forthcoming from the ministry. Ashe dismantled the telescope and suspended operations in 1873 to make way for a new observatory building. That seems to have spelled the end to Ashe's solar research. Before the 1882 transit of Venus, he had withdrawn from active work. Retiring in 1883, he removed to Lennoxville and died in Sherbrooke in 1895.

Ashe has been overlooked, yet, in a century of practical astronomy, he is the pioneer in astrophysical research in Canada. He had hoped to make Canada's name known in the world of astronomy, but he had no immediate successor. Under later directors at Quebec – Lieut Andrew Gordon, RN;

Ashe's son, William Austin Ashe; and Arthur Smith – the observatory reverted to the routine efforts intended of it by the government.

During the time of Ashe's work at Quebec, state-supported Canadian astronomy moved into a second phase, one of co-ordination and consolidation. In the early 1850s there were three observatories in what would become Canada – at Toronto, Quebec, and Fredericton. The first was an imperial establishment with little interest or equipment for astronomy, the second a Canadian institution, and the third a university facility. They were unconnected, unco-ordinated. Yet, within twenty years, a small network of observatories was linked together under the authority of the minister of marine and fisheries in Ottawa and under the direction of the Meteorological Service of Canada. An auspicious beginning for government-backed astronomy turned out to be a failure for astronomy – though a great success for meteorology – and was eventually displaced by another branch of government science. None the less, the men of this first network were the primary practitioners of astronomy in Canada until the turn of the century. Beginning with Toronto and Quebec, the network added observatories at Kingston, Montreal, and Saint John.

In the 1850s, Kingston was a bustling port town, a one-time capital of the Province of Canada, the home of a garrison of some size, and the site of Queen's University. Although small, Kingston had the ingredients for at least a limited intellectual society, and astronomy was of interest to several residents. James Williamson (1806–1895), professor of mathematics and natural philosophy at Queen's, had made some observations to try to determine the longitude of Kingston by means of lunar distances and eclipses. A published account of his work had come to the attention of Thomas Devine of the Crown Lands Office of Canada West, who was constructing a map and had found Bayfield's longitude of Kingston too far west.[55] At the time, Williamson possessed a 3-foot refractor, a clock, a sextant, and a theodolite,[56] but no observatory. An able amateur appeared in 1854 – Lt-Col Baron de Rottenburg, who possessed a Dollond telescope of 2.5-inch aperture and 3.25-foot focal length. On 26 May 1854, an annular eclipse of the sun was observed by Rottenberg and others. He had, somewhat earlier, attempted to obtain a new equatorial from a New York manufacturer, but as it was not forthcoming, he went to see Alvan Clark in Boston, who was willing to provide a 6.25-inch equatorial for $800. With local interest piqued by the eclipse, a committee was established to investigate purchasing a Clark telescope and erecting it in a suitable building. The committee – Rottenburg, Williamson, a Mr Rowan, a Mr Taylor,

RN, Judge Burrowes, and Dr Horatio Yates – reported to the city in 1855 that a site in the new park was suitable.

The telescope was fully subscribed; it arrived in the autumn of 1855 and was erected in a small tower in the park in the spring of 1856; it lacked a clock drive, and there was no transit instrument or clock.[57] The people of Kingston had access to it, however, and in 1858 Williamson observed Donati's Comet, his account appearing in the *Canadian Journal*.[58]

A grant of $500 was obtained in 1860 and again in 1861 from the provincial government to match grants given to Dr Charles Smallwood in Montreal. It was clear that arrangements were inadequate, and so in 1861 moves were made to enlarge the observatory. The observatory was deeded to Queen's University by the city in 1861.[59] William Leitch, the principal and one-time assistant to Nichol at Glasgow Observatory, met G.P. Bond on a visit to Harvard and requested information concerning the addition of transit and computing rooms to the existing tower.[60] Bond sent Williamson a lengthy letter describing the arrangements in Cambridge and offering suggestions about the size, shape, and construction of the building and on the mounting of the transit and clocks. He also suggested instruments that should be purchased, from his brother's firm, and that the plans for the Kingston Observatory ought to be drawn up by his brother-in-law.[61] Williamson replied that the committee had made its own plans, including a transit room as large as Harvard's, but that he wanted to order a micrometer eyepiece and illuminator from Clark to make the telescope more useful.[62]

Leitch, writing from Europe, mentioned that he had purchased a stellar catalogue, was sending over a telescope for the college, and would also give his own clock to the observatory.[63] The board of visitors – Mayor Overton Gildersleeve, Edward Berry, Yates, and Burrowes – saw that there were insufficient funds for a proper observatory. In 1862, three petitions, to the legislative assembly, to the legislative council, and to Gov-Gen Viscount Monck, requested further funds. Despite two grants and money raised by public subscription, the board was still in debt $500 and wanted more: 'Mr. Airy, the Astronomer Royal who had been corresponded with on the subject, informs them [the Board] a Transit Circle costing together with an Astronomical Clock not less than £650 Sterling still requires to be provided in order to entitle the Observatory to rank as a Scientific and National Institution.'[64]

Although this additional funding was not obtained, the $500 grant became annual; it was still $500 in the 1890s. Since the board could not purchase a good transit, it next requested, through the governor-general, that the RAS be approached concerning the gift of an instrument.[65] An old transit by Cary was sent out in 1864 – a gift to the society in 1829 by Beaufoy and called the Beaufoy Transit – as a more-or-less permanent loan.[66] In addition, a small Simms transit was purchased for $180. Now an observing assistant would be required, but, as the board reminded itself in June 1863, the salary would be $500 per annum, requiring the entire government grant, while the debt was still $500; none the less, it decided to try to raise money by subscription to hire a Mr Cross as observer.[67] Cross evidently did not last long, for on 2 December Leitch reported to the Queen's trustees that Nathan Fellowes Dupuis had been hired as observer at £55 per annum.[68]

The engagement of Dupuis was an excellent idea. Born near Kingston in 1836 of a French-Canadian father and Loyalist mother, he learned clock- and watch-making and later turned to private tutoring in mathematics. In the year of his employment, he entered Queen's and he was graduated with first class honours in 1866; he took an MA in 1868. From 1864 to 1868 he acted as observer, then took up the chairs of chemistry (1868) and mathematics (1880) at Queen's.[69] His talent as a clockmaker was a godsend to the impecunious observatory. He was sent to Harvard by Williamson in 1864 to be shown the working methods by Bond.[70] The mean-time clock given the observatory by Leitch passed back to his estate in 1864, and so Dupuis built a replacement as a companion to the sidereal clock he had already constructed. With the Beaufoy transit mounted and the Simms transit in use for zenith observations, Williamson and Dupuis made observations of occultations, Jupiter's satellites, and the latitude of Kingston.[71]

After Dupuis had taken up his duties as chemistry professor in 1868, the work fell to Williamson, whose interests were meteorological. Williamson had married Sir John A. Macdonald's sister, Margaret, but in his extant correspondence with Sir John, there is only one mention of the observatory: in 1868, he inquired whether the annual grant was to be provincial or federal and added that Airy had sent out a set of Greenwich observations.[72] By then Kingston had joined the small network of Dominion observatories.

The origins and growth of the next link in the chain, McGill Observatory, differ in detail from those of Kingston, but both arose out of amateur interests and ended up as part of the government network. It seems curious that Kingston should have acted before Montreal, the metropolis of Canada. While medicine and natural history flourished in Montreal, the physical sciences numbered few adepts. This changed with the appointment of John William (later Sir William) Dawson (1820–1899) as principal of McGill College in 1855. Dawson was already a geologist and palaeontologist of some note and soon brought those sciences into the forefront. Dawson resuscitated McGill's arts and science faculty and encouraged astronomy and meteorology.

The beginnings of the McGill Observatory were not in Montreal but nine miles west in St-Martin, Île Jésus, where, in the 1840s, the meteorological observatory of Dr Charles Smallwood was erected.[73] Smallwood was a remarkable figure; although he never became a professional in the sense that E.D. Ashe was, his contributions went far beyond that of contemporary Canadian amateurs. Born in Birmingham in 1812, a graduate in medicine at University College, London, he left for Canada in 1833. He took up the study of meteorology; over the years his meteorological and astronomical observations were printed in the local papers and in the *British-American Journal*. He contributed papers on ozone, snow crystals, and other meteorological and astronomical topics to *Silliman's Journal* in the United States and to the *Canadian Journal* and *Canadian Naturalist*. His instruments were mostly home-built. His 1858 article in the *Canadian Journal* gives a description of his observatory.[74] Of astronomical interest, the ridge of the roof had a slit and shutter for transit observations, a transit being mounted on the floor in the centre of the observatory. In addition, Smallwood had a Fraunhofer-built 7-inch achromatic telescope of 11-foot focal length. It was mounted equatorially with setting circles, though it was portable and had to be carried outdoors for observation. With it he observed comets, eclipses, and other phenomena. It was probably through Dawson's influence that Smallwood was named unpaid professor of meteorology at McGill in 1856.[75] Two years later, E.T. Blackwell, president of the Grand Trunk Railway, 'proposed to erect an astronomical Observatory if sufficient means and suitable site could be obtained,' and Dawson suggested the university campus.[76] This observatory was intended to provide a time service for the railway. Nothing came of the proposal for the moment. In the mean time, Smallwood desired a government grant to finance his observational work, for which he applied in January 1859. This was refused, but he was able to have the bishop of Montreal and other influential friends lobby the government.[77] Later that year he received $500, and the grant was repeated the following year.

In July 1862, Smallwood wrote the board of governors at McGill proposing to move his apparatus to the campus if a suitable building were provided. This request was received favourably by the board.[78] He then wrote the government apprising it of his intention.[79] In August plans for a stone building were accepted, and in October a committee reported an estimate of $1,945; the board agreed to proceed.[80] The stone building on Carleton Road was ready for occupancy in 1863. Fortunately, the annual

grant to Smallwood was not withdrawn, and the new observatory continued to receive $500 annually. The initial instruments were presumably the
transit and refractor from Île Jésus Observatory. Smallwood made no
mention of a clock at Île Jésus, so a clock must have been purchased after
the new observatory opened, since, by 1870, the facility was making its
practical contribution by providing time to the city, harbour, and government buildings in Ottawa via telegraph. McGill might have had an
observatory seven years earlier had the Grand Trunk – which could have
afforded it – followed up its idea; the state, of course, through its sub-
ventions to Smallwood, became the real founder of the observatory.

Smallwood died in December 1873 and was replaced by Clement Henry
('Bunty') McLeod (1851–1917), a native of Cape Breton and student of
Smallwood's at McGill. McLeod was a civil engineering student and received his BA in the year of Smallwood's death. His post was that of
observer in meteorology, not professor, a post which had lapsed. Unlike
Smallwood, McLeod needed a salary, and the board hoped for more government money to provide for meteorological observations.[81] The time
service at McGill ended at this point and was transferred to the private
observatory of Charles Blackman, a prominent businessman.

Before Confederation, the observatories at Toronto, Quebec, Kingston,
and Montreal were voted operating expenses by the provincial legislature;
Toronto and Quebec were entirely funded by government, and Kingston
and Montreal each received $500 annually. In 1867, funding passed to
federal hands. Initially nothing changed. But in September 1868 Peter
Mitchell wrote Lord Monck requesting, on the advice of the government
auditor, that the Quebec Observatory be turned over to his department,
since it 'is specially designed to give the time to the shipping.'[82] Two
proposed observatories, for Saint John and Halifax, for similar purposes,
ought also to come under his department. In 1869, Prof Kingston at Toronto
and Sandford Fleming (1827–1915), the chief engineer of the Intercolonial
Railroad, suggested that the observatories be united for meteorological
reporting.[83] In July 1871 money was set aside for a Meteorological Service,
under the direction of Kingston, but no attempt was made to create such
a network until 1874, when the new minister, Albert Smith, requested the
privy council to put all the observatories under his control.[84] This was
done.

The fifth observatory of the network was at Saint John, NB. Built on
Fort Howe Hill in 1870, it was equipped with a time-ball with modifications
suggested by Ashe[85] and a transit instrument to correct the clock. While
under the direction of George Hutchinson, it burned in 1877 and was
replaced in 1881 with a new building nearer the harbour. Since the instal-

lation was small and meant only to serve as a time and weather station, no research was ever undertaken there. The Department of Marine had intended to build a similar post in Halifax, and although some funds were voted for it, nothing was ever accomplished.

During the 1870s, astronomy was cultivated very little throughout the new network. Prof Kingston remarked in 1875 that the Toronto Observatory 'is not furnished with apparatus suited for astronomical researches. Our astronomical observations are not made in the interests of astronomy, but are subservient to other purposes, and are almost entirely confined to transits for time.'[86] Williamson had little to report from Kingston, apart from the usual meteorological records. In 1881, Queen's built a new observatory on the campus, and the land of the former building was deeded back to the city.[87] This was ultimately replaced by another structure on campus. Smallwood was dead by 1873, McLeod involved only in meteorology, and Hutchinson at Saint John attended to his routine duties. Ashe, with his observatory temporarily out of commission, had ceased active research with his solar photography. Time was given to Saint John, Quebec, Kingston, Montreal, and Toronto by their respective observatories; Montreal also provided time to Ottawa by means of telegraph.

From Smallwood's death in 1873, until 1879, time for Montreal was provided not by the McGill Observatory but by businessman Charles Blackman. In 1879, he moved to New Haven, Conn, and donated his instruments to the college observatory. These included a 7-foot equatorial with a 6.25-inch objective by Henry Fitz; a 3.25-inch-aperture, 42-inch-focal-length transit instrument by Jones; a Howard mean-time clock; and a sidereal clock.[88] The McGill tower was rebuilt and a dome added to accommodate the new equatorial. With these instruments in place, McGill took up the time service once again.

The great astronomical event of the next decade was the transit of Venus in 1882, a chance to determine the distance to the sun. The coming event was the first in Canada since 1769, as the transit of 1874 was not visible to North America. Most of the observatories were by then under full or partial control of the Canadian Meteorological Service; the new director, Charles Carpmael (1846–1894), a Cambridge graduate, was enthusiastic about observing the transit. Carpmael had been Kingston's deputy since 1872 and succeeded him in 1880 on the latter's retirement. Carpmael's lobbying with his minister was successful: Parliament granted $5,000 in May 1882, and the director was asked to co-ordinate the work.[89] The extent of Carpmael's plans was evidently not public knowledge; in a letter to the Toronto *Mail* of 20 May, the amateur astronomer and writer Mungo Turnbull explained to the public the value of the transit and suggested that observing sites be chosen from Labrador in the east to Woodstock in the west. He then wrote Sir John Carling asking the government to appropriate sufficient funds to observe the transit and was referred to Prime Minister Macdonald, who also held the Interior portfolio. In a letter to Macdonald,[90] Turnbull argued that Canada required a proper astronomical observatory so that longitudes could be measured from a Canadian standard meridian. He mistakenly supposed that Canadian longitudes were all taken from maps but nevertheless believed that the approaching transit was an excellent opportunity for building such an observatory. The idea of a national observatory was a little premature, and Macdonald was the last politician to approach on that subject. In any event, Carpmael had already organized the project.

During the summer of 1882, Carpmael sent his assistant, Lieut Andrew Gordon, RN, to England to discuss the best observing methods. He returned with four Admiralty chronometers and a model for practice at the telescope. In September Carpmael and Professors Johnson, McLeod, and G.H. Chandler of McGill and Brydone Jack from Fredericton practised with the McGill refractor. Another session was held in Toronto for the Toronto observers – Prof Williamson from Queen's, Wolverton from Woodstock, and others.

At McGill, Alexander Johnson (1830–1913), professor of mathematics, had been quietly lobbying for a new observatory since 1878. He wrote the local papers early in 1879 after holding public meetings. One response came in the form of a gift of the Blackman instruments. In addition, a 4.25-inch refractor was given him by the Trafalgar Institute, and a 4-inch refractor was loaned by G.A. Drummond. When the government grant was received in 1882, McLeod elected to go to Winnipeg, taking the first telescope; F.R. Blake, DLS, and B.C. Webber, of the Meteorological Service, were deputized to observe in Ottawa with the Drummond refractor and the McGill portable transit.

Carpmael, at Toronto, had spent £400 from the government grant on a 6-inch Cooke refractor, the first large telescope possessed by the Magnetic and Meteorological Observatory. At Quebec, Lieut Gordon was to observe with the Clark refractor, assisted by W.A. Ashe. Since the event was of interest to the public, Ashe wrote to Mgr T.-E. Hamel, rector of Laval, inviting anyone who cared to observe with them.[91] Other observers were chosen across the country: A. Allison, in Halifax, with a 4-inch Dolland refractor; H.A. Cundall, in Charlottetown, with a 4-inch refractor; Brydone Jack, in Fredericton, with his 6-inch Merz; T.S.H. Shearman, in Belleville, with a 4-inch refractor; Prof Abram Bain, at Victoria College in Cobourg, with a 4.25-inch refractor; Prof Williamson, in Kingston, with the 6.25-inch Clark and the Beaufoy transit; Prof Hare, at Ladies' College,

Whitby, with a 6-inch Fitz refractor; and Prof Newton Wolverton, at Woodstock, with Woodstock College's 8-inch Fitz refractor, stopped down to 6 inches. Virtually every major telescope in Canada was pointed at the sun on 6 December. The weather was such that only the Ottawa and Kingston observers obtained even three contacts, Winnipeg two, and Belleville and Cobourg one each.

Before and after the transit, time signals were exchanged between Toronto and the stations by telegraph. The few Canadian observations made little impact. The 1874 transit had been observed by many nations in hopes of being able to measure the solar parallax – the angle subtended by the earth's radius at the earth-sun distance – but had not proved very successful. The British results, both visual and photographic, yielded values from 8″.08 to 8″.88, an unsatisfactorily wide range. As a result, the British were less involved with the 1882 transit.[92] Nonetheless, the Canadian results were included in the final British report.

The 1880s and early 1890s were relatively quiet for Canadian astronomy. Once it had resumed giving time, the McGill Observatory increased its activities, providing time not only to the city but also to Ottawa and the railways and, by telegraph, to Bermuda, Jamaica, and the Azores.[93] Timeball operations continued at Quebec and Saint John. Time was given by the Toronto Observatory, but the new Cooke refractor seems to have been generally idle until R.F. Stupart, Carpmael's successor, inaugurated daily observations of sunspots in 1895. The most energetic worker of the time was C.H. McLeod. In June 1883, he and W.A. Rogers of the Harvard College Observatory exchanged clock signals by telegraph to redetermine the longitude of the observatory, correcting the 1857 value of Ashe, made from Viger Square, not the McGill campus.[94] In 1888, McLeod began observing sunspots with the Blackman telescope, employing ruled paper discs to locate the spots by heliographic longitude and latitude. The first series (1888–90) was published in 1890; a second continued observations to the spring of 1891.[95] From his records, McLeod found that the sunspot minimum occurred at a different time than stated by E.W. Maunder, who headed the Greenwich Observatory solar program. McLeod observed the spots visually, while Maunder employed a photoheliograph. McLeod seems to have dropped solar work at this point; it was not taken up again in Canada until J.S. Plaskett's work some fifteen years later.

One aspect of government astronomy in this period, patronage, seems alien to us now. Although a Civil Service Act had been brought in as early as 1875, many appointments were still in politicians' hands. University observatories such as McGill and Kingston were largely immune, despite their governmental connections. But the Quebec Observatory provides us with an interesting political history. E.D. Ashe wished to retire from the directorship in 1882; even before he retired, a ship's master, Capt Duttan, wrote Sir John A. Macdonald asking for the post. Apart from being an amateur meteorologist, he had 'taken over the Canadian Mails for 25 years and never lost a bag' and 'the Canadian government owes me some slight acknowledgement in some situation.'[96]

Carpmael, who had a poor relationship with Ashe, as their bristly correspondence shows, had kept the observatory on a short leash financially. On Ashe's retirement in May 1883, Carpmael sent Lieut Gordon back to Quebec to take charge; Gordon removed the instruments from the Bonner Farm observatory and placed them back in the Citadel, in the hands of the army. But Gordon was called away in 1884 to participate in the British scientific expedition to the Arctic, and a new observer was needed. In January 1886, William Austin Ashe, PLS, DLS, received his father's old post and returned the instruments to the Bonner site, but his dealings with Carpmael were not much warmer than his father's. He was well qualified for the work but died prematurely in 1893.

A scramble began anew. The priests at the Séminaire de Québec watched knowingly: '[M. O'Leary] fait des demarches pour être chargé de l'observatoire de Québec ... La Nomination est une affaire exclusivement politique. Le succès sera à celui qui deploie la plus grand somme d'influence.'[97] O'Leary evidently did not. In March 1894, the post went to Arthur Smith, son of a prominent Conservative. His only background was surveying, which was sufficient to operate the time service and take meteorological readings. In Quebec, the majority felt that the proper candidate was A.-P. Roy, a professor at the Collège de Lévis, who had references from such notables as Camille Flammarion. The French-language press in Quebec was incensed, l'Electeur claiming that Torontonian Carpmael could not know whom to appoint in Quebec and was simply handing out patronage.[98] By the following year, Roy had worked out a new scheme: to divide the observatory at Quebec into two sections – meteorological and astronomical – the latter for himself. This he proposed to Prime Minister Laurier in 1897.[99]

Rumours swirled about: Smith would be removed and replaced by Roy; Smith had made the observatory a secret meeting place for les bleus, and so forth. Roy had already enlisted the aid of P. Langelier, MP, who replied that he had little power to grant patronage.[100] The minister of marine and fisheries, Louis Davies, refused to divide the position, on the advice of the new head of the Meteorological Service, Frederick Stupart (1857–1940). This, to Roy, was treachery from a Liberal government, but he assumed that it was Stupart's doing, since he was 'un tory – probablement un fanatique.'[101] Smith's appointment may well have been simple patronage, but it is to Laurier's credit that he did not follow up on precedent and remove him for a Liberal. (Smith discharged the routine work of the observatory until 1929, when he was replaced by Maurice Royer.) By the time of these patronage squabbles, the second phase of government astronomy had already played itself out. The third phase, which crystallized into the permanent form of state-supported astronomy, had come to fruition.

At the time of the transit of Venus, government astronomy was fully in the hands of the minister of marine and fisheries and the Canadian Meteorological Service. The best instruments were in their observatories in Quebec, Montreal, and Toronto. However, a rival group was soon to be formed which, in less than twenty years, would come to dominate government astronomy. This group, formed of surveyors in the Department of the Interior, had its origins in Canada's 1869 acquisition of Rupert's Land, the Hudson's Bay Company's vast territories in the Northwest. With the formation of Manitoba in 1870, the entry of British Columbia in 1871, and the Macdonald government's promise in 1871 to build a transcontinental railway, opening the land to settlers became urgent. Accurate land surveys were undertaken by the federal government from 1869. To better organize the opening of the west, Macdonald formed the Department of the Interior in 1872, with himself as minister. The surveying was accomplished under the surveyor-general's scrutiny. Astronomical work done by this branch was in connection only with surveying, not with timekeeping or any other practical needs.

Astronomy had already reached the west with the International Boundary Survey of 1872, which surveyed the 49th parallel. Although it was a joint American-British venture, Canadians were included, notably William F. King (1854–1916), destined to become the head of the new group of government astronomers.[102] King, born in England, had spent most of his life in Canada, being educated in Port Hope, Ont., and at the University of Toronto, where he took an honours degree in 1875 after his return from the Northwest. He became both a dominion land surveyor and a topographical surveyor in 1876 and returned to the prairies the following year. By 1881, he was an inspector of surveys. This work was not essentially astronomical, and the Department of the Interior had taken no overt interest in the subject. Astronomy became important in 1885, for surveying the 'railway belt' – an area along the Canadian Pacific line in British Columbia, twenty miles to each side of the line. Since the terrain was mountainous and could not be surveyed in the traditional way, locations had to be fixed astronomically. This task was entrusted to Otto Julius Klotz (1852–1923), son of a German immigrant to Ontario. Klotz took his BA at the University of Michigan in 1872. After several years of private and contract surveying, he was appointed 'Astronomer' in charge of the railway project, the first civil servant with that title.

A third notable surveyor became part of this nuclear group, Edouard-Gaston-Daniel Deville (1849–1924). Born in France, he studied at the Naval School in Brest and was attached to hydrographic surveys in the French navy until his retirement in 1874.[103] In 1873 and 1874, he worked with E.D. Ashe in determining longitudes along the Ottawa River. After executing various surveying commissions for the Quebec government, he entered the service of the Interior Department in 1880. In the following year, he became an inspector of surveys. His rapid rise led in 1885 to his appointment as surveyor-general of Canada. He was elevated to director-general of surveys in 1922. Deville had both a practical and a theoretical background in astronomy and had written a practical book upon his arrival in Canada;[104] he also wrote with King in 1881 a manual for surveyors. As a founding member of the Royal Society of Canada, he contributed several papers on practical astronomy to its *Transactions*. He was in charge of surveying in Canada when Otto Klotz began the railway belt survey in 1885.

The astronomical group in the survey consisted of Klotz and Thomas Drummond in the field from 1885, with King as supervisor.[105] The group began in Seattle, as there was no western Canadian site with telegraphic connections to an observatory, and moved eastwards. It soon became evident that new instruments were required and that a permanent observatory to aid in longitude determinations of western points was desired. This was the practical need not yet demonstrated in Turnbull's letter to Macdonald some years earlier. King accordingly sent a memorandum to the minister, Thomas White, in February 1887, asking that a small observatory be built in Ottawa. This was approved. King and Klotz visited the United States the following month to inspect instruments. A Cooke transit was ordered that summer, and the instruments already in their possession were placed in a temporary observatory in King's Ottawa garden. Word of the proposed observatory leaked to Charles Carpmael in Toronto, and he complained to his minister that the Toronto and Montreal observatories in his charge were adequately performing the tasks envisioned for the new observatory. This may account for Klotz's remark that White told King in October 1887 that he had been opposed in cabinet and decided to drop the matter.[106] The opposition must have been voiced by George Foster, minister of marine.

In the following year King and Klotz enlisted the aid of Deville, who in turn impressed A.M. Burgess, deputy minister of the interior, with the necessity of an observatory. The project was stalled again with the death of White in April. He was succeeded by Edgar Dewdney. Burgess convinced Dewdney that the astronomical work of the ministry ought to have a permanent head and that King should become chief clerk, with the title chief astronomer. This would mean a salary of $1,800 per annum, for which money was set aside in the departmental estimates for 1889–90 and 1890–1, but the post was not created. On reception of another memorandum from Burgess, Dewdney wrote to the Privy Council in June 1890 requesting the change.[107] This was acceded to a week later by an order-in-council of 30 June 1890.

With Prime Minister Macdonald's death a year later, there followed a succession of short-lived Conservative governments, and the observatory project was shelved at the ministry. Boundary surveys and a new determination of the longitude of the McGill Observatory linked telegraphically with Greenwich occupied the department's astronomers. The fortunes of the group changed abruptly with the election of Wilfrid Laurier's Liberals in 1896. The new minister of the interior was the energetic and influential Clifford Sifton. King and Sifton had a congenial relationship from the beginning. By 1898, the observatory project was alive again: after receiving a memorandum on the subject from King, Sifton inquired how many telescopes were available in Canada. King replied that there were only two – the 6-inch refractors at Toronto and McGill – though it was hoped that Sir William Macdonald might provide a new instrument for McGill.[108] That he overlooked the Quebec Observatory's 8-inch telescope, which had performed pioneering work, is mute testimony to how political intrigue and official neglect had nearly wiped out the usefulness of that institution.

King wanted a 10-inch refractor with a spectrograph for astrophysical research. To complement the practical instruments, housed since 1890 in a temporary wooden observatory on Cliff Street, near the present Supreme Court building, new clocks, a pendulum, and a chronograph were to be ordered. Once Sifton agreed with the proposal, events proceeded rapidly. By May 1899, the Ministry of Public Works had placed $16,000 in the supplementary estimates for 1900.

The next two years were spent hammering out details. King and Sifton wanted a site near the Parliament Buildings, but Klotz argued that it was unsuitable. A too-ornate plan provided by the office of the chief architect was turned down and another substituted. Meanwhile, King began negotiations with Warner and Swasey in Cleveland for the equatorial. He became convinced that a 15-inch refractor would be more suitable; Sifton agreed, and the $14,600 instrument, with optics by Brashear, was ordered in June 1901.[109] About this time Sifton decided upon the Central Experimental Farm as a better site, and King and Klotz readily agreed. The telescope and other equipment arrived during 1904, and the building was in use by April 1905. The Dominion Observatory, Canada's first national astronomical institution, had become a reality. The final cost had been $310,000, a phenomenal amount for science at that time.

King was able to obtain not only a fully equipped observatory but also sufficient staff.[110] Several officers came from the Astronomical Branch: surveyors under King (all temporary staff on annual contracts) included J.J. McArthur, C.A. Bigger, E.T. deCoeli, J.M. Bates, T.A. Davies, N.J. Ogilvie, and F.W.O. Werry. 'Inside workers' were W.M. Tobey and F.A. McDiarmid (both computers), J.D. Wallis (photographer), and L. Gauthier (draughtsman). Two recent Toronto graduates who were to become leading figures at the observatory, Robert Meldrum Stewart (1878–1954) and John Stanley Plaskett (1856–1941), were retained as a computer and mechanician, respectively.

Plaskett had been mechanical assistant at the Toronto physics laboratory and was involved in photographic research when he approached King for a position.[111] One of Plaskett's teachers, Alfred Baker, wrote to Sifton, who in turn told King that he 'would favour this man's appointment.'[112] However, the only permanent staff members were King and Klotz; the others could be lost at any time. The work of the observatory, described in chapter 4, began before completion of the building.

A period of some sixty years is covered in this chapter. At the beginning, about 1840, Canada was a collection of colonies with small populations, few financial resources, and little economic development, virtual wards of the politicians at Westminster and the bureaucrats at Whitehall, both anxious to rid themselves of Canadian responsibilities. By the turn of the twentieth century, some five million Canadians had begun to flex their economic and political muscles, and, while not yet a fully independent nation, their country was a magnet for immigrants and investors who knew of its economic potential. Before the 1840s, the practical benefits of astronomy were secured by the state – the imperial government. Most practitioners were not Canadian, and local interest in the science was rare and sporadic. After 1840, as Canada slowly began to take shape politically and economically, the practical benefits of astronomy became, increasingly, an object of local government concern. We pass, then, from astronomy in Canada to Canadian astronomy.

This Canadian astronomy naturally had similarities to that in the neighbouring United States, also a young, developing country, but vast differences in population and in land and economic resources meant that the science would diverge considerably in the two countries. The Canadian attitude about government intervention ensured that most astronomical activity would be state-supported. When the state's needs were still small, and its resources circumscribed, little activity was funded, and we see the isolated surveys of Jack, Ashe, or the western exploring parties. The grants to small observatories in Quebec, Kingston, and Montreal were intended to extract only practical information of immediate value. The pragmatic ideology of Victorian Canadians, coupled with small-scale government, meant that almost no resources could be applied to pure research, to generate knowledge about the universe. The solar work of E.D. Ashe – which ended prematurely for lack of funds – was one of the few examples of pure research in the period. The observations of the transit of Venus in 1882 were equally non-practical, but by the 1880s Canada was beginning to think of itself as a nation, and participation in an international effort was a good advertisement of maturity. In that sense, the investment of $5,000 for the transit observations was relatively cheap, although even that sum was not easily wrested from Ottawa.

The first generation of astronomers was not born or educated in Canada: Williamson and Jack were Scots, and Ashe, Smallwood, and Bayfield, English. By the 1870s, men born or raised in Canada – McLeod, W.A. Ashe, Dupuis, King, and Klotz – began to replace them. The most marked feature of Canadian astronomy was its very limited manpower. There were never more than a half-dozen astronomically trained men in government service until almost the end of the century. Still, Canada's needs in practical astronomy were not great and required neither many men nor many elaborate institutions. Canada had no empire and required no Greenwich; it had no large military and merchant navy and thus required no US Naval Observatory. Railways and increased shipping activity underscored the need for the provision of time, while western expansion necessitated longitudes and latitudes. The state needed no more and provided no more.

CHAPTER THREE

Students and Public

Astronomy's place in the curriculum of Canadian universities has two functions: the intellectual broadening of the liberal arts student and the professional training of the astronomer. The former is a notable feature of nineteenth-century higher education in Canada – indeed, to a much greater extent than today – but professional training is essentially a twentieth-century story and will be deferred to parts III and IV. Astronomy found at least a small place in the curriculum at nearly all colleges and universities, where similarities among courses, especially in smaller institutions, are striking. Therefore, this survey will provide detail on a few representative patterns at the Séminaire de Québec (later Université Laval), the University of New Brunswick, Queen's University, McGill University, and the University of Toronto.[1]

Astronomy was not, in the nineteenth century, a subject for primary or secondary schools and was found exclusively in colleges and universities. Higher education was distinctly Canadian, whether francophone or anglophone, though its roots and influences were a mixture of French, English, Scottish, Irish, and American educational practices and outlooks. The Oxford-Cambridge classical tradition did not take root for a variety of social, political, and economic reasons. The Scottish model – more practically oriented than the English – together with certain American innovations and, later, transposed German ideas, brought about a unique blend suitable for a small northern country. Science and mathematics were integral parts of the curricula in nearly all institutions of higher learning; in some schools, the arts curriculum was fully half science and mathematics, though by the end of the century the move towards specialization had begun and the retreat from those subjects began in earnest.

The academics of the time believed that science had a place in the curriculum for three major reasons. First, science and mathematics encouraged the student to think logically. Second, it showed the student the ways in which God created and regulated the universe; this view was consistent with the eighteenth-century passion for natural theology, the reconciliation between the Bible and science. Third, it gave the student a grounding for more practical subjects such as engineering, agriculture, or medicine. Although the curricula touched upon nearly all the sciences, including astronomy, courses in virtually all institutions were superficial and elementary. Few universities had proper libraries, museums, or observatories. Until late in the century, laboratories were non-existent. Canadian schools were not consciously involved in the active training of scientists, though many scientists were graduated from Canadian schools. The graduate school, the doctoral degree, and individual research would not come until the 1890s and then only at large schools such as McGill and Toronto. The PH D in astronomy is a mid-twentieth-century phenomenon in Canada.

There were two ways in which astronomy was presented. In one format, practical work was included, occurring when a school had instruments or an observatory and a professor interested in practical applications. The other method was purely descriptive, well suited for small schools. In some cases, the armchair approach might be made more rigorous – as at Toronto – by the addition of mathematical astronomy, often drawn directly from Newton's *Principia*; this approach was the more typical.

As we will see, the teachers of astronomy were almost never professional, or even amateur, astronomers themselves. In most universities, the professor of mathematics and natural philosophy (physics) taught astronomy. In some cases, an engineering professor might handle the courses. In the collèges classiques of Quebec, astronomy was normally detailed to the general science teacher. The first generation of such teachers, except in the Quebec colleges, was not Canadian either by birth or by education, but rather British and, even more specifically, Scottish. As the century progressed, more Canadian-born, Canadian-educated professors replaced the immigrant teachers. This pattern seems to hold for most of the sciences in Canada.

Although nineteenth-century Canadian astronomy was dominated by governmental activities, the ranks of amateurs continued to grow, particularly in the cities. The small size of the professional community resulted in an early association between it and the amateurs – another distinctive feature of Canadian astronomy. Astronomers were often isolated and were typically associated with the local scientific or literary societies for intellectual contact. This form of association continues with the present-day Royal Astronomical Society of Canada (RASC). There is, presumably, a critical mass of professionals that must be reached before the disengagement from amateurs begins to take place. In Britain, the RAS catered to both amateurs and professionals in the nineteenth century but eventually emerged as strictly a professional organization, with the British Astronomical Association and other groups absorbing the amateurs. In the United States, the Astronomical and Astrophysical Society of America (renamed the American Astronomical Society) was formed in the late nineteenth century strictly for professionals, and it has remained so. Local societies, and later specialist societies and the Astronomical League, were formed by amateurs. In Canada, this division did not come until the 1970s, with the formation of the Canadian Astronomical Society, and the disengagement of professionals from the RASC has yet to take place. Among Western nations, Canada seems to be unique in this regard. The size of Canada's population, and its distribution, seem the primary determinants of this pattern.

With the cessation of teaching, in 1757, at the Jesuit Collège de Québec because of the war with Britain, higher education ceased to exist north of New England. Any rudimentary teaching of astronomy certainly ended in Quebec when Bonnécamps departed in 1759. The early years of the British régime were not altogether propitious for a renewal of higher education, but the Roman Catholics of Quebec, without Jesuits, were allowed to reopen the Petit Séminaire de Québec in 1765. The Séminaire, founded by Bishop Laval in 1668 as a residence for the theological students of the Collège de Québec, took up the teaching role of the now-defunct Jesuit institution and soon was offering the full classical course to a small number of students, both French and English.[2]

By the turn of the nineteenth century, the Séminaire provided the full course which typified the collège classique: a seven-year program which was more than secondary education but less than a university degree. The last two years, known as the philosophy course, included science and mathematics. Astronomy was part of this course. Antoine-Bernardin Robert (1756–1826) taught philosophy from 1790 to 1795. Miscellaneous notes in the archives of the Séminaire written about 1790 by Robert include questions and problems on eclipses and descriptions of gnomons and sundials and their construction.[3] Another notebook from about the same time has calculations for the solar eclipse of 4 July 1796.[4] Robert was probably the first teacher at the Séminaire to teach much astronomy, although there is in the archives a curious statement of account dated June 1770 from the carpenter, Pierre Emond, for work on a 'chambre d'observasion.'[5]

Robert was succeeded in the philosophy course by Jérôme Demers (1774–

1853), who had been taught by Quebec surveyor Jeremiah McCarthy. Demers taught science sporadically for forty years, and his interest in astronomy garnered him the title of 'l'homme qui lit dans les astres' by the citizens of Quebec.[6] A notebook from 1816 describes the calendar, motions of the sun and planets, eclipses, and longitude and latitude of the moon, with problems requiring the use of the *Nautical Almanac*.[7] The notes of Demers's course taken by abbé Montminy in 1820[8] give a clear idea of the nature of astronomy teaching at the Séminaire. The basic concepts of the celestial sphere and the planets were introduced: these included seven known planets, including the planet 'Herschel,' and four minor planets. Kepler's laws and Newtonian mechanics were discussed, and stellar astronomy was up to date, with mention of William Herschel's ideas on the distribution of stars. There follow the other stock topics: comets, eclipses, tides, sundials, the calendar, and numerical problems relating to these.

The astronomy part of the philosophy course was taught largely without instruments until the arrival of abbé John Holmes (1799–1852).[9] Holmes was an American convert to Catholicism and one of the leading forces for better science in Quebec schools. By the 1830s, the Séminaire de Québec had been joined by seven other collèges classiques in the province – at Montreal, Nicolet, Ste-Hyacinthe, Ste-Anne-de-la-Pocatière, Ste-Thérèse-de-Blainville, L'Assomption, and Chambly. They all offered much the same course of studies. Holmes aided not only the Quebec college but others as well, and in 1836 he travelled to Britain and France to purchase scientific equipment for them. He called on George Airy, the astronomer royal, for advice on purchases.[10] Nicolet received a Dollond achromatic of 3.5-foot focal length, as did Ste-Anne; Quebec, a Troughton and Simms theodolite.[11] Holmes wanted an observatory for his own college and noted: 'L'établissement n'a point encore d'observatoire, quoiqu'il se soit déjà procuré plusieurs beaux instruments d'astronomie.'[12] Holmes's sister, also a convert, had become a nun and teacher at the Ursuline Convent School in Quebec. While in France, Holmes procured for the convent a celestial globe and a beautiful orrery, both still in the convent museum. Thus, it is clear that astronomy was taught to francophone girls in the 1830s, but it is unlikely that such fine demonstration devices were available to the few schools for anglophone girls of the time.

The astronomical part of the collège classique course in philosophy had its public side, for the examination of the boys was an annual event, entailing disputations on a problem chosen by the teacher. Astronomy was taught by a succession of well-known teachers, including Louis-Jacques Casault (1808–1862), a student of Demers and later first rector of Laval, and E.-A. Taschereau (1820–1898), the first Canadian cardinal and a student of Holmes. Casault increased the amount and quality of science teaching at the Séminaire, and, thanks to him, astronomy was made a separate professorship in 1843, the first such chair in Canada.

In 1852, the Université Laval was created by charter, with Casault as its head. Unlike English-language universities, it did not initially offer courses for the BA degree but examined students who had completed the classical course in the colleges. If they were successful, they received a Laval BA. When courses were introduced, their function was to prepare students for the examination. Astronomy never became an important part of the curriculum during the century. No observatory was immediately built for the university, although by the time of Confederation Laval possessed the best university library and finest science museums in Canada. The relative position of astronomy can be seen in the Laval *Annuaire* for 1876–7, in which the science requirement for the BA was set at 200 lessons in mathematics, 160 in physics, and 40 in astronomy.

Not everyone in Quebec was happy with the situation of the sciences. Thomas-Etienne Hamel (1830–1913), a rector of Laval, who studied physics and astronomy in Paris, argued to the Royal Society of Canada during his presidency: 'But are educated youth in our country in a position to take part in these serious studies, and to advance science, not simply as a work of the imagination, but as a careful study of the facts, in order to draw rigorous conclusions from them? Alas! I can clearly see obstacles in their path. Allow me to point out three of them, that I would gladly compare with the plagues of Egypt, at least where they concern French Canada. They are journalism, the civil service, and politics.'[13] These professions, long so attractive to young French Canadians, continued to be so for decades to come, and the place of astronomy changed little during the remainder of the nineteenth century. The Séminaire finally obtained an observatory in 1892, installed on its roof.[14] It was available to the university, as the institutions shared the same campus and many of the same faculty members. The telescope augmented the lectures, and occasionally professors wrote their own textbooks. The best known of these was the *Cours élémentaire de cosmographie* (1913), by Henri Simard (1869–1927), who taught astronomy at the Séminaire from about 1893 to 1923. Although the pattern of education in Quebec differed in many respects from that in other provinces, the raison d'être for teaching astronomy was essentially the same: to enrich the student's education. It was never meant to be a stepping-stone to a scientific career. Even so, very few students completed the philosophy course in the classical colleges, and fewer still went on to become professional scientists. None became professional astronomers.

The earliest English-speaking colleges in what was to become Canada sprang up in the Maritime provinces,[15] beginning with the Anglican King's College in Windsor, NS, which began regular, degree-granting operations in 1807. A similar institution, founded also by United Empire Loyalists, was King's College in Fredericton, NB, which became a full-scale college in 1828.[16] These two institutions created two of the earliest observatories in Canada. Other Maritime colleges, including Acadia, Mt Allison, and Dalhousie, taught astronomy as part of the liberal arts curriculum, but in none of them did astronomy flourish during the nineteenth century.

The arts curriculum of the small Maritime colleges in the early part of the century included the classics, some theology, history, a little mathematics, and perhaps some natural history. These early curricula were patterned on the Oxford-Cambridge tradition, and at contemporary Oxford and Cambridge science was at a low ebb. In Canada, the arrival of Scottish-trained professors from the British schools where science was cultivated made a distinct difference. This was true at both King's colleges. There was no distinct chair of astronomy in anglophone schools. Typically, a college had two science professors – one for chemistry and natural science, which included geology, and the other for mathematics and natural philosophy or physics. The latter usually had charge of astronomy.

According to Bishop,[17] the college at Windsor had obtained astronomical instruments in 1810 and in 1842, implying that some astronomy was part of the natural philosophy or mathematics course. In 1855, Rev John Hensley, styled professor of mathematics, natural philosophy, and astronomy, was teaching the rudiments of the science. He was succeeded in 1859 by Joseph David Everett (1831–1904), a Glasgow graduate. In 1860 Everett began to interest the governors of the college in building an observatory. After some delay, the building was completed in 1862. It was equipped with standard surveying instruments, a Ramsden altazimuth, transit, sextant, and artificial horizon, as well as a telescope manufactured by the London firm of Cary.

Everett's astronomy course, offered to seniors, was, judging by the calendar entry of 1863, of a practical nature. It included spherical trigonometry, the celestial sphere, the use of globes, and basic ideas in surveying and navigation.[18] To aid with the work he had obtained an orrery and globes. While all liberal arts students were required to take the astronomy course, it neither accounted for a great deal of time in the curriculum nor went beyond the fundamentals of the science.

On Everett's departure in 1864, J.E. Oram, educated in Ireland, succeeded him. Oram was styled professor of mathematics, natural philosophy, astronomy, and engineering – which underscores one of the occupational hazards of teaching in a small college. His contemporary at Fredericton,

Loring Woart Bailey, appropriately described his chair as a 'settee.' Oram employed Galbraith and Haughton's popular text on astronomy as preparation for the baccalaureate examination. On his retirement in 1880, astronomy, as a full course, disappeared. By 1903, the only astronomy shows up in the fourth-year physics course by B SC students. After Oram's departure, the observatory fell into disuse; it was eventually pulled down, and the fate of the telescope is unknown. The only work of note done there was meteorological and, excepting the recent work by Bishop, has escaped the notice of historians altogether.

In Frederiction, astronomy had a longer, though chequered, career.[19] King's College, like its sister institution in Windsor, was an Anglican foundation but had the good fortune to hire Scottish professors of science. The first two of these, James Robb, M D (1815–1861), and David Gray (c. 1811–1856), were appointed to the college in 1837, the former in chemistry and natural history, the latter in mathematics and natural philosophy. We do not know whether Gray taught astronomy, as the first calendar did not appear until 1861. Gray departed in 1840 and was replaced by William Brydone Jack (1819–1886), who had studied under David Brewster at St Andrew's. Jack and Robb found natural allies in one another for the furtherance of scientific education on the Scottish model. In 1847, they wrote a joint letter to the college council requesting money for instruments; their rationale neatly encapsulates the meaning of science education for many nineteenth-century Canadian professors:

It is vain to found Chairs of Mechanical Philosophy or of the History of Nature, if the means of properly illustrating the same are to be withheld; it is vain to institute courses of Astronomy or Optics if the realities of the Telescope and the Microscope cannot be exhibited ... if the true mastery which mind has acquired over Matter is to be clearly apprehended – if the glorious triumphs of Art in this 19th Century are to be realized by the Youth of New Brunswick, we do not see how it can be done without a great and early effort to procure an adequate stock of Philosophical Apparatus.[20]

There follows a request for £1,000.

In this letter is one of the two themes of nineteenth-century science in Canada – man's mastery of nature – which was repeated continually on both sides of the Atlantic in the heyday of the Industrial Revolution. Astronomy was an excellent example of how the human mind can penetrate beyond the bounds of the earth. The other theme is touched upon by Robb in his Encaenial address of 1839: 'A double advantage will result from the study of [science], if the teacher strives constantly to impress upon the young men committed to his charge the necessity connecting means with ends, and then again with their final purpose in Creation, and the intrinsic imperishable evidence which they afford of care, divine superintendence, and special providence.'[21] Any professor in any British North American college could have written these words.

The request fell on receptive ears; the usually impecunious college was able to set aside £300 for the 'mathematical department' for the purpose of obtaining 'a good 7 feet Achromatic Telescope and such other apparatus as may be more immediately necessary.'[22] This was the essential first step towards obtaining the observatory, but in the mean time Robb was making inquiries of his own. He was an amateur meteorologist who had given a course of lectures on the atmosphere at the Saint John Mechanics' Institute in 1842, and he wrote to Col Edward Sabine, then director of the magnetic observatories of the empire, to obtain for Fredericton a magnetic and meteorological observatory like Toronto's. Lt-Gov Sir William Colebrooke had passed on to Robb copies of the reports from Greenwich and Toronto for his perusal; Robb wrote him on 12 November 1846 suggesting that such an observatory be built in Fredericton.[23] Colebrooke passed the letter along to Earl Grey, colonial secretary, who in turn passed it to Sabine. The latter was enthusiastic and urged that the £5,000 necessary be set aside.[24] There were good arguments for having astronomical, magnetic, and meteorological observations made in Frederiction, chief among them settlement of land title disputes. This sounds like a dubious argument, but surveying without a well-determined meridian was a continuing difficulty, leading, a few years later, to Jack's longitude work and his assistance to surveyors. The college had even offered the site.[25] Cost presumably militated against the proposal, and the whole issue died quietly in the spring of 1847.

This ended the idea of a government observatory attached to the college, but not the observatory project. Jack had communicated with the Munich firm of Merz, successors to Fraunhofer, and had ordered a telescope. Thereupon, the college council got cold feet and resolved on 5 April 1848 that:

the telescope now proposed by Professor Jack at an original cost of £336 Sterling with the building and apparatus necessary for its proper use – are more than the wants of the College require or the funds will warrant. That ... [a] committee of this board authorized to make arrangements for annulling the order given for such a Telescope on the best terms in their Power and to secure the best instrument which can be obtained at a charge not exceeding £100 Sterling over and above the sume of £110 Sterling already advanced to the manufacturer.[26]

It appears that Messrs Merz were able to comply, for an order to pay the firm an additional £100 was made in October.[27] The actual building of the observatory – to Jack's specifications – did not occur until the winter of 1850–1 and was completed by March 1851, after the expenditure of £170.[28] The small, wooden building with its exquisite brass and mahogany telescope still stands on the University of New Brusnwick campus, having been renovated and named a national historic site at the instigation of J.E. Kennedy and others in the 1950s. Jack himself was proud of the wooden structure, with a wooden dome in the shape of a truncated cone, and its attached transit room. The telescope was not quite what he had hoped for, as he wrote Sir George Airy: 'It was a mere mercantile transaction and the instrument forwarded without examination, it is not so perfect as it should be, there being several minute air bubbles in the Object glass, and the chromatic aberration imperfectly corrected. It has hitherto been used more for gratification of the curiosity of the students and others than for any scientific purpose.'[29]

With the opening of the observatory, Jack was able to modify his course in astronomy. Although there was no calendar before 1861, we know the content of his course for 1853, thanks to a memorandum he sent to the council.[30] This memorandum was one result of a protracted debate in New Brunswick concerning the usefulness of King's College. The Anglicans wished the college to maintain its theological and classical character, but more forward-looking people wanted practical education available to all, not just to the élite. Jack and Robb, as good Scottish-trained professors, appreciated this outlook and attempted to inject what practicality they could. The debate came to a head in 1859, when the legislature secularized the institution. In the following year, Jack became president of the renamed University of New Brunswick. With this development, the future of practical eduation was assured.

Jack's 1853 memorandum – written to assure the harried council that his courses had practical elements – outlines the astronomy component: principles of optics, orbits, moon, sun, and planets, basic notions of celestial mechanics, the use of instruments, eclipses, time and calendar, tides, comets, fixed stars, variable stars, multiple stars, nebulae, stellar motions and distances, the nebular hypothesis, and system of the world. This was a broad, though elementary survey but necessary, for, as Jack comments: 'As Professor of Mathematics and Natural Philosophy, Astronomy cannot strictly speaking, be said to fall within my province; but several years ago I felt myself constrained to undertake it, as I was unwilling that our young men should leave College, without acquiring some knowledge of this the most perfect of all the physical sciences.'[31] With the addition of instruments and an observatory, the course could become even more practical. Although the time allotted for astronomy was slight, because New Brunswick offered a three-year course, all students were required to study it.

The 1861 calendar shows that astronomy was offered in the third or senior year and had as texts Robert Main's *Rudimentary Astronomy* (London, 1852) and Galbraith and Haughton's popular *Manual of Astronomy* (London, 1852). Students wishing honours certificates could read, in addition, Elias Loomis's *Introduction to Practical Astronomy* (New York, 1855), which describes, among other topics, the exchange of telegraphic signals for longitude purposes, a technique pioneered by Jack in British North America in the 1850s. The observatory's instruments were available for practical astronomical and civil engineering instruction.

Jack retired in 1885, and his successor, Thomas Harrison, brought the university in line with other schools with a four-year degree. As no one with Jack's interest in astronomy occupied the physics or mathematics chairs, the observatory fell into disuse. Astronomy ceased to be part of the curriculum for all students.

At Mt Allison College, a small Methodist school founded in 1858 in Sackville, NB, astronomy was part of the regular BA course. Small and poor, the college enrolled only 107 students in 1891, but they were given astronomy in the fourth year of their BA or PH B studies. By 1905 astronomy was reduced to just a part of fourth-year mathematics, taught by Prof Sydney Hunton. A school with a similar history is Acadia University, founded by Baptists at Wolfville, NS, in 1839. The calendar for 1861–2 notes that Daniel Higgins, professor of mathematics and chemistry, taught nautical astronomy to second-year arts students. Higgins taught astronomy until 1891, but the course underwent a variety of changes throughout the period, given alone, as navigation, or not at all. In 1891, Frank R. Haley was appointed alumni professor of physics and astronomy, but he did not change the course, which, in 1893, was made an elective instead of a requirement. By 1899, the course was no longer allowed even as an elective, being only one hour of lectures weekly. The course was elementary: the required texts were Ball's *Elements of Astronomy* and Young's *General Astronomy*. The calendars of the 1890s speak of observatory work.

The centre for science in the Maritimes was Dalhousie University, which opened in Halifax in 1863. At first it followed the pattern of the other Maritime schools. The mathematics professor, Rev Charles Macdonald, taught astronomy during the summer session, but it was not a regular part of the winter curriculum. The sciences grew rapidly in the 1890s, largely thanks to the philanthropy of expatriate businessman George Munro.

Unfortunately, as the calendars chronicle, astronomy was not included in this largesse. It was missing in the 1890s and was, by 1904, only part of the surveying and geodesy course for civil engineering students.

In central Canada, where institutions were somewhat larger than their Maritime counterparts, the needs were similar and educational patterns much the same. McGill College in Montreal, although founded in the 1820s, was not a fully functioning university until the appointment in 1855 of John William (later Sir William) Dawson as principal. Dawson was a Nova Scotian, already a noted geologist and a one-time student at Edinburgh; he intended from the outset to give science an important place in the arts curriculum. During his administration – he retired in 1893 – astronomy was taught on a similar plane to that of other Canadian colleges. By the turn of the twentieth century, traditional astronomy had begun to wane at McGill; to this day the university has no astronomy department.

McGill's curriculum in Dawson's early days resembled New Brunswick's, though within a four-year framework. Astronomy came in the third year as part of the natural philosophy course. From 1857, this was taught by Alexander Johnson, trained in mathematics at Trinity College, Dublin. The calendar for 1864–5 denotes the course as 'Mathematical Physics and Astronomy,' the astronomical texts being Galbraith and Haughton's *Manual of Astronomy*. Students wishing to read for honours would be examined on Hymer's *Astronomy* and Sir John Herschel's *Outlines*. There was no observatory, although some of the calendars refer to the course as one dealing with the practical aspects of the subject, and we might infer from this that the celestial sphere, navigation, and time and calendar were emphasized, but that observation was minimal. The honours degree requirements were stiffer: the calendar for 1870–1 notes that a student must pass an examination based upon Main's *Rudimentary Astronomy*, Godfray's *Lunar Theory*, and Newton's *Principia*. This was, however, only for students in the mathematics and physics option.

The prospects of astronomy at McGill were seemingly improved by the same means as that for other sciences: private philanthropy. Just as Macdonald and Redpath money aided the other sciences, so the gifts of two dedicated Montreal-area amateurs, Dr Charles Smallwood and Charles S. Blackman, aided astronomy. Smallwood's observatory, it will be recalled, was moved to the McGill campus from St-Martin in 1862. The Blackman instruments were added in 1879. Smallwood had been associated with the university for some time; in 1856 he had been appointed unpaid professor of meteorology. Despite his interest in and knowledge of astronomy, the course continued under Johnson's control. The observatory had, by 1880, an excellent complement of instruments but did not become a centre of astronomical education. The calendars of this period state that students could receive instruction in the use of meteorological instruments from Smallwood, and later from C.H. McLeod, but they remain silent on whether astronomy students could avail themselves of the astronomical instruments.

The time service initiated by Smallwood and carried on by McLeod and the meteorological work which continues to this day reflect the priorities of McGill just before the turn of the century. Engineering was growing rapidly, and the earth sciences – Dawson's special interest – made use of the observatory, but a trained astronomer was neither on staff nor hired. By the turn of the century, just after Dawson's death, McGill had become an important scientific and engineering centre, but astronomy did not enjoy the growth that physics and chemistry did. In 1900–1, when the curriculum was revised, a half-course in descriptive astronomy was available to third- or fourth-year students in the mathematical and physical sciences. Honours students could take a fourth-year course in physical astronomy which concentrated on celestial mechanics. There was no longer mention of observatory work.

Astronomy appears as part of the liberal arts curriculum at most of the colleges and universities of Ontario in the nineteenth century. Queen's College was founded in 1841 by Scottish Presbyterians of Kingston. James Williamson (1806–1895), born in Edinburgh and educated at its university, joined the faculty in 1842 as professor of mathematics and natural philosophy. The post had already been offered to William Brydone Jack at Fredericton, who declined it on account of the weakness of the scientific content of the proposed curriculum.[32] Williamson was a beloved figure in his old age, but, as Calvin remarks, 'though Glasgow made him LLD in 1855, he is not specially remembered as scholar or teacher.'[33] His part in the observatory project has been recounted; his publications were few.

The Queen's records show that up to about 1850, astronomy was not taught in either the mathematics or natural philosophy courses. In 1851–2, Herschel's *Outlines* was made part of the course in natural philosophy,[34] but it soon dropped out. In the first published calendar (1861–2), the natural philosophy course is described as including Herschel and the first three sections of Newton's *Principia*. The 1863–4 calendar states that Galbraith and Haughton's textbook was used and that the observatory was available for teaching. It is curious that astronomy had not appeared earlier in the Queen's curriculum, since the Reverend Principal William Leitch had a great interest in the science, owned a telescope, and had been instrumental in having the Kingston Observatory built. Some of Williamson's undated lecture notes in astronomy have been preserved: they are very general, though there are some comments on the use of the spectroscope.[35] Williamson's successor, N.F. Dupuis,[36] was much more enterprising and energetic. Not only did he construct the clocks used in the observatory, he also built two orreries for instructional use. His interests were inclined more towards mathematics than towards astronomy, but he wrote two textbooks on the subject: *Elements of Astronomy* (1910) and *Spherical Trigonometry and Astronomy* (1911). He had one notable student in astronomy, S.A. Mitchell, who graduated in 1894 and went on to be a professor at the University of Virginia and director of the Leander McCormick Observatory. After Dupuis's death in 1917, astronomy re-entered the doldrums at Queen's and stayed there for many years. As at McGill, energies seemed to be directed into other sciences and into engineering, despite possession of an adequate observatory for teaching.

The University of Toronto had a stormy early history.[37] Founded in 1827 by John Strachan as King's College, strictly Anglican like its namesakes in Windsor and Fredericton, it did not open its doors until 1843. Although its curriculum owed much to Dublin and its faculty was largely English, the program evolved into the Canadian style of liberal arts eduation, with a large portion of the course work given over to science, some of it practical. In the early years, although the Toronto Observatory was adjacent to the campus, astronomy was not even a full-fledged course of its own but was taught as part of natural philosophy. James Loudon (1841–1916) recalled that in 1860 'Astronomy was propounded with the aid of drawing on a blackened globe, some spheres of brass that represented the sun and moon and stars, and the endless formulas that filled the board.'[38] The instructor was J.B. Cherriman, the mathematics professor. He was also director of the Magnetic Observatory until G.T. Kingston's arrival.

Astronomy was simply part of the mathematics course and not a subject in its own right. By 1889–90, astronomy was reserved for fourth-year honours students in mathematics. The mathematics professor, Alfred Baker, employed Chauvenet's *Spherical Astronomy*. By 1905, astronomy was divided into two courses in a manner similar to that at McGill. An elementary course was avilable, and a few students might elect the advanced course, which concentrated on celestial mechanics, with texts by Godfray and Chauvenet, plus the occasional use of the Toronto Observatory. The fortunes of astronomy at Toronto changed soon afterwards with provision for instruction in astrophysics by Clarence Augustus Chant (1865–1956), but the groundwork was laid earlier. In 1875, Loudon, Cherriman's student, followed him as professor of physics. He was keen to build up a physical laboratory and modernize the physics curriculum. After some resistance – the Oxford-Cambridge style of education was still preferred by many at the university – he succeeded.[39] Loudon later became president of the university and fostered the introduction of the PH D for science. Chant was one of his students; J.S. Plaskett was his mechanical assistant in the 1890s. Chant was appointed a lecturer in physics in 1892, but despite his great interest in astronomy that subect was still in the hands of the mathematics department.

Since 1892, Victoria College had been federated with Toronto. Victoria's history dates from 1836, when the Methodists, led by Egerton Ryerson, had opened the Upper Canada Academy in Cobourg, Ont. Both boys and girls attended and were taught astronomy, because Ryerson was a firm believer in science in secondary and tertiary education. As he pointed out in his inaugural address as first president of Victoria in 1842: 'In this Province ... students have acquired little or no knowledge of Natural Philosophy, Chemistry, Mineralogy, Geology, Astronomy, etc., except what they have obtained in another Province, or in a foreign country. If one branch of education must be omitted, surely the knowledge of the laws of the universe, and the works of God, is of more practical advantage, socially and morally, than the knowledge of Greek and Latin.'[40] From that time until federation with Toronto, Victoria had a professor of mathematics and astronomy, first William Paddock, then William Kingston, and from 1870, Abram R. Bain. Bain was something of a polymath, already an outstanding classics scholar before his appointment. In 1868 he studied with Benjamin Peirce at Harvard and, in 1869, with Chauvenet in Paris.[41] The calendar for 1875 typifies the program: astronomy, surveying, and navigation were offered in the fourth year to all BA students, since practical science was then considered the best course of studies. Sometime before 1882, English friends of Bain's provided the college with a 4.25-inch refractor by Smith, Beck and Beck of London. This was mounted in a revolving turret on the Faraday Science Hall and employed for observations of the transit of Venus. At federation, science teaching passed to Toronto, and Bain was named professor of ancient history.

We may briefly note three other small schools. Western University, now the University of Western Ontario, at London provided one course in descriptive astronomy for students in the 1880s and 1890s but little else in astronomy until the 1920s. Manitoba, the only university west of Ontario before the turn of the century, offered no courses at all in astronomy but allowed students to write examinations in the subject.

Astronomy did thrive for a short time at Woodstock College, a small Baptist foundation in Woodstock, Ont.[42] A small observatory was erected in 1878–9 by Prof Jabez Montgomery, but Newton Wolverton, professor of mathematics and principal 1881–6, put it on a proper footing with an 8-inch Fitz refractor, a 2-inch transit instrument, and a chronometer. Wolverton was involved in the 1882 transit on Venus observations, but he resigned from the college in 1886. From 1896 T.S.H. Shearman was in charge; he engaged in solar researches and had ambitious plans for photometric studies. In connection with the latter he began construction on an 18-inch reflector. Shearman later went to British Columbia to work for the Meteorological Service. By 1910, the observatory had ceased to function; the college itself was absorbed by McMaster University in 1926 and disappeared.

We tend to forget, in these days of highly specialized and professionalized science, the cultural values and intellectual entertainment afforded by scientific subjects. The modern reader is often surprised to discover the extent of the general interest in science during the nineteenth century. At that time, there were no electronic media and very little else to occupy one's mind except reading, hearing lectures, and exchanging ideas with like-minded people. Education was not yet taken for granted, for many men and women had come to Canada or even grown up in the country with little or no formal education. None the less, the thirst for knowledge was great, and the belief – today slightly disreputable – that humans had and could master nature through science and technology was almost universally held.

The great passion of the nineteenth-century public was the earth sciences: geology, mineralogy, geomagnetism, palaeontology, and meteorology. Astronomy, related as it is to these subjects, was also in vogue – indeed, probably more than at any time in the present century, save until very recently. The public had a variety of means by which to learn more about the heavens. Reading was the simplest route. Few public libraries existed, fewer still had good science collections, and good bookshops were scarce, but popular British and American astronomy books were available and were read. Unfortunately, there are no statistics on the readership and availability of books, though surviving library catalogues indicate modest astronomical holdings. With the rise of the mechanics' institutes in the 1830s – designed to raise the educational levels of workmen, or 'mechanics' – libraries and reading-rooms became available to many more people of diverse social standing. Public lectures were a typical feature in such institutes, particularly in Ontario, where local governments provided financial assistance. Public lectures were also sponsored by schools, colleges, universities, and churches.

For those wanting more intellectual stimulation, local literary or scientific societies filled the void; however, only a few existed – mostly in cities and larger towns – and they usually required a nucleus of university professors to sustain the intellectual content. Astronomy was not a great interest of any of these societies, though astronomical papers were occasionally read at their meetings and published in their proceedings. Typically, these societies had a very small group of professional scientists together with a much larger group of amateurs, dilettantes, and those merely wishing for genteel society. The only group that approached the status of a truly professional scientific society, the Royal Society of Canada, was not founded until 1882. Although it numbered nearly all the important professional astronomers as members, the subject was scarcely noticeable amid all the geology, engineering, and natural science. Strictly astronomical societies do not make a permanent appearance until the 1890s.

Scientific societies and public libraries did not appear in Canada until the nineteenth century. There was no scientific society in New France, though Galisonnière corresponded with several scientifically inclined officers and physicians and was himself a member of the Académie royale des sciences. There was no public library in New France. Little changed with the coming of the British, but, by the early nineteenth century, small groups such as agricultural, literary, and library associations had arisen in Quebec and Montreal. The first society in Canada to take any interest in science was the Literary and Historical Society of Quebec (LHSQ), founded in 1824 by Lord Dalhousie.[43] Despite its name, its members' interest in science was manifest from the outset, and both Dalhousie and the legislature provided funds for scientific instruments. The society was for the upper class and professionals of Quebec. A more popular group, Société pour l'encouragement des sciences et des arts au Canada, was formed in the city in 1827. The two were amalgamated in 1829 through the exertions of Lt-Gov Sir James Kempt. The new LHSQ had about one-quarter francophone membership. Capt Bayfield was a member and contributed papers to the meetings. In 1829 the society inaugurated its *Transactions*, further volumes appearing irregularly throughout the century. The Rebellion of 1837 seriously crippled the society; by the 1840s, when the seat of government had been removed to Kingston, the society had dwindled to a handful of members. It revived in the 1860s, when Lieut Ashe was an active member.

The *Transactions* printed some of the earliest astronomical articles in Canada. In the early nineteenth century Isaac Watts was observing sunspots from the Citadel, and these were recorded in an early volume of the *Transactions*. The early years of the society were devoted more to geology, natural history, and meteorology, but there were a few astronomical talks given to the society and a few papers published. In 1837, Mr Justice John Fletcher, who had an interest in mathematics, published a set of rules for ascertaining longitude by the lunar distance method.[44] A young Englishman, Valentine

Daintry (d. 1842), reworked Fletcher's rules with spherical trigonometry.[45] This probably represented about the highest mathematical attainment of anyone in Canada at the time.

The society's renewal in the late 1850s and early 1860s also witnessed the work of Lieut Ashe, who provided a number of papers to the *Transactions* and read other papers to the meetings. G.T. Kingston, who taught navigation to pilots at Trinity House in Quebec before becoming director of the Toronto Observatory, spoke to the society on aerolites. James Douglas jr, a prominent member and son of the famous physician of the same name, kept the society informed about astronomical spectroscopy, although he himself undertook no active research. Towards the end of the century the LHSQ reverted to its historical mission, and little more science was included in its transactions, although there is a delightful and, to the modern reader, very patronizing lecture to women on the nebular hypothesis by William Austin Ashe.[46]

Interest in astronomy does not seem to have been as great in French Canada as in English Canada. Yet several people in Quebec followed it; aurorae were of interest, and reports appeared in Michel Bibaud's *Bibliothèque canadien* in the 1820s. Indeed, the French-language press always seemed to carry more scientific news than the English press. In Quebec, the Université Laval was the centre of astronomy. With the installation of a reflecting telescope on its roof in 1892, local amateurs were able to view occasionally. The inauguration of the telescope was held in conjunction with the partial solar eclipse of 20 October 1892, and 'la plupart des prêtres, quelques écclesiastiques et les physiciens sont montés.'[47] There were, however, no astronomical societies in French Canada.

During the nineteenth century, most Canadians lived in small towns or on farms. They had little recourse to societies for intellectual stimulation and had to rely on newspapers, journals, and libraries, especially those of the mechanics' institutes. Astronomical books were found in some of these libraries, but catalogues show that they were few in number, often out of date, and rarely circulated. This was true even in the cities. The best library in Quebec was that of the legislative assembly, which was removed to Kingston at Union in 1841. Its catalogue shows that there were several popular elementary books on astronomy, but nothing for those wishing something more serious.

For towns where public or college libraries were non-existent or small and societies unavailable, the local centre of intellectual stimulation was the mechanics' institute. These institutes grew up in Canada during the 1830s, being modelled on the institutes in Great Britain. George Birkbeck of Glasgow, the originator of these institutions early in the century, intended them to be places where working people could improve their minds and morals. The typical Canadian mechanics' institute had a library and reading-room, and sponsored lectures or other entertainments. Some attempted to offer night courses for tradesmen. However, as in Britain, these institutions were taken over by the middle class and run more for its benefit; industrial workers and artisans do not seem to have readily taken these institutions to their bosoms.

Lectures on astronomy were given in a number of the mechanics' institutes, Toronto and Preston, to name two. One complete set of such lectures survives: the 1857 astronomical lectures to the Saint John Mechanics' Institute by William Brydone Jack.[48] Based largely on Sir George Airy's popular *Ipswich Lectures*, Jack's lectures brought basic information to an audience that had no local university or society. Jack lectured in Fredericton, as well, and astronomy must have come up occasionally at the meetings of the Fredericton Athenaeum, which, though small, was active during the 1840s and 1850s, thanks to the efforts of James Robb.

By the early 1850s, there were thirty-four mechanics' institutes in Canada West, but only five in Canada East, where they were dominated by anglophones. The largest was, of course, in Toronto; it was founded in 1831, one year after Quebec's. Its annual reports from 1846 onwards tell the tale of a society that had an important place in the intellectual life of the city until about 1860, after which its size remained relatively steady while the city around it grew rapidly. Public lectures were given to large audiences. In 1846, of eleven lectures, two were on astronomy, one on comets and the other on the lunar theory by a Mr Dwyer. Thomas Henning of the Canadian Institute lectured on astronomy in 1852 and on the plurality of worlds in 1855. By 1860, however, public lectures ceased altogether, owing to the growing competition of the Canadian Institute and the University of Toronto. The library remained, but science was not a large part of it. In 1868, of some 7,400 books in its possession, only about 600 are listed as being on 'science and art,' with science books the least circulated, compared to the extremely high turnover of novels. In a city of approximately 50,000 inhabitants, those seriously interested in astronomy were few: in 1868, at the founding meeting of the Toronto Astronomical Club, only eight turned out.

Local scientific societies were later in developing in Canada West. The largest city, Toronto, had to make do with the Mechanics' Institute as a focal point for intellectual stimulation until 1851, which saw the re-establishment of the Canadian (later Royal Canadian) Institute. For anyone with an interest in science, Toronto seemed a wasteland until that time. Lieut Riddell, in charge of the Toronto Magnetic Observatory in 1840, wrote plaintively to Edward Sabine: 'I hope you will fulfill your promise of coming out here this year or the next. I have not found anyone here who appears to take the least interest in the operations or on whom I could rely for the least assistance. There is a college and three or four Professors but I have only seen one of them, the shallowest little wretch I ever met – my only chance is in the arrival of some officer who may have a turn to science.'[49]

An attempt had been made in 1849 to form the Canadian Institute as a professional organization for engineers, surveyors, and architects, but the response was exceedingly poor and the new society nearly collapsed. However, in 1851 the society's purview was enlarged to include science and technology, and the membership was thrown open. Receiving a royal charter in November 1851, the new society elected William Logan, director of the Geological Survey, as president, with Capt Lefroy as first vice-president. Within a year, it boasted 175 members, drawn not only from Toronto, but from across the province and nearly 20 from Canada East. Among the members were several with a strong interest in astronomy: Capt Lefroy, Prof Cherriman, Rev G.C. Irving, and Dr Smallwood of Montreal.

In 1852, the Institute inaugurated the *Canadian Journal of Industry, Science and Art*, an excellent publication that compared favourably with the *Scientific American*. It carried, in addition to institute news, original articles by its members, reviews, news of other societies, and the latest scientific intelligence from abroad. Astronomy figured largely from the first, with many news items, reports from the Greenwich Observatory, and reviews of astronomical books.

The institute took a leading role in 1853 in persuading the provincial government to take over the Toronto Observatory, while the *Canadian Journal* kept its readers abreast of its activities; Cherriman, director of the observatory, reported his meteorological data and observations of meteor showers. In 1854, in preparation for the annular eclipse of that year, Cherriman and Irving were asked by the institute to prepare a memorandum on how to observe the eclipse.[50] On the day of the eclipse, 26 May 1854, Cherriman and Irving observed with three others at Prescott with a 4-inch Gregorian reflector, a 2.75-inch refractor, and a chronometer. This eclipse attracted a good deal of attention: the Toronto Observatory staff observed with a small refractor, the Montreal observers were rained out, and Col de Rottenburg and F.J. Rowan observed from Kingston. The primary work done by these teams was meteorological rather than astronomical. Henry H. Croft, chemistry professor at the University of Toronto, watching the eclipse with a small telescope in his yard, noticed that his chickens went to roost during the eclipse, though the sky did not seem to darken perceptibly, and that the number of flies seemed to diminish.[51]

Rottenburg became a member of the institute in the mid-1850s. An active amateur astronomer, he was well read and lectured to the institute in 1856 on the supposed self-luminosity of Neptune.[52] In the following year, he reported on the occultation of Spica, a curious observation, since he noted that all the bright stars the night of the occultation seemed planetary in appearance. During the occultation, it seemed to him that Spica was within the limb of the moon, then disappeared abruptly. He adds that a Mr Chalmers, an RAS fellow residing in Barrie, had observed the same thing with a 3.5-inch Dollond refractor. Rottenburg argued that this observation might give some credence to the idea of a lunar atmosphere, an idea still alive among some European astronomers.[53] Rottenburg's last contribution to the *Journal* was a log of observations of sunspots from January to March 1858.[54] Accompanied by interesting drawings, the article offers nothing more than a description of his observations; he did not seem to have followed up on this line of work.

One other member of this period should be noted: Thomas Henning. Little is known of him, except that he was active in both the mechanics' institute and the Canadian Institute in the 1850s. He was very well versed in the latest fact and theory and read papers on the asteroid belt and on meteors.[55]

During the 1860s, the *Canadian Journal* shrank somewhat and its likeness to *Scientific American* diminished. Apart from a few items on astronomy, the *Journal* had no articles or Canadian reports until about 1880, when Sandford Fleming, a charter member, roused the society with his proposal for standard time.

Astronomy figured in the writings of Canadian natural theologians. Natural theology, an attempt to explain religion by means of science, was an important intellectual current of the eighteenth century, especially in Britain. As the triumphs of science of the nineteenth century increased in number, theologians were pressed to reconcile the new geology, palaeontology, and astronomy with the biblical account of creation. One of the best-known Canadian books, James Bovell's *Natural Theology*, was published in Toronto in 1859, the year Darwin's *Origin of Species* appeared. There were many other such works: Sir William Dawson's later books of this nature were a rearguard action against Darwinism.

Two Canadian Studies are of interest. In the early 1820s, Henry Taylor of Quebec devised a scheme to reconcile the Genesis account with the latest findings of chemistry, geology, and astronomy in *A System of the Creation of Our Globe*, which, by 1854, had gone through nine editions. It is calm and erudite – Taylor obviously kept abreast of science with the latest books and British reviews – and was well received in Canada. A more curious work, by Ezekiel Stone Wiggins, appeared in 1864. *The Architecture of the Heavens*, like Taylor's *System*, shows a good knowledge of contemporary astronomy but, as with the books of many enthusiastic but not quite knowledgeable enough amateur scientists, comes close to being in the class of crackpot ideas. Wiggins's tone is much more combative than Taylor's; he particularly ridicules the notion of William Leitch, Queen's principal and avid amateur astronomer, that the sun is fire and comets provide fuel for the sun. Wiggins had been reading Sir John Herschel and agrees that the sun was gaseous but then proffers a curious theory that the solar atmosphere must be electricity. This, he believes, is shown by the fact that comets' tails are repelled when near the sun. Wiggins's idea on the source of comets is stranger still. For contemporary astronomers, the nature of sunspots was a great mystery, but it was evident to Wiggins that they must represent solar volcanoes that eject comets by their repulsive electrical charges.

Neither Taylor nor Wiggins was truly a popularizer of astronomy. In fact, very few popular accounts have ever been penned in Canada; two early examples might be noted in passing. William Leitch published *God's Glory in the Heavens* in 1862. It is ironic that Leitch, who was so conversant with astronomy, should have left the teaching of the subject at Queen's to James Williamson, who seemed to have little gift for popularizing it. A somewhat curious contemporary was Henry Hayden, an Irish-born Anglican cleric who learned astronomy from the celebrated Brinkley at Trinity College, Dublin. His life was unstable, but he did write a series of popular columns on astronomy for a Halifax newspaper in 1836; the articles were collected into a small book, *Illustrations of Astronomy*, the same year.[56]

The Canadian scientific community was small and scattered in 1882 when the Marquis of Lorne addressed the first meeting of the new Royal Society of Canada.[57] He had worked actively with Sir William Dawson of McGill to form a society not unlike the venerable namesake institution in London. It was to be both honorific and active in its pursuit of knowledge. Of course, Canada could not sustain a fully scientific society like London's, and so an organization composed of four sections – French letters, English letters, physical sciences, and natural sciences – was proposed instead. The new society, at least in its honorific sense, was relatively successful from the beginning. Dawson, Canada's best-known scientist, was its first president; P.-J.-O. Chauveau, former Quebec premier and educational advocate, served as the first vice-president. Among the twenty members of section III (physical sciences) were Alexander Johnson, Charles Carpmael, E.G.D. Deville, Nathan Dupuis, and Sandford Fleming.

Over the next few years, three astronomers, King, Klotz, and McLeod, were elected, as places fell vacant through death or retirement. The *Transactions* published original research from the beginning. Non-members, too, could publish articles submitted by a fellow. Articles by both McLeod and King appeared in this way. Fleming, Carpmael, Deville, Johnson, and Dupuis all published astronomical items in the pages of the *Transactions* in its first years. Curiously, after about 1896, astronomy articles dropped out of the *Transactions*; in fact, section III had always been the weakest section in the early years. After the turn of the century, there were other outlets for astronomical research results. Staff members of the new Dominion Observatory published in the *Report of the Chief Astronomer*, while members of the Astronomical and Physical Society of Toronto had their *Transactions*. Some, like J.S. Plaskett, published in the new American *Astrophysical Journal*.

As a forum for professional astronomy, the Royal Society of Canada was a failure. There were simply too few astronomer members. The society did confer upon those members a feeling of élan, as any organization claiming to be composed of the best can, yet only a truly scientific society could be an agent of the professionalization of science – in this case, the feeling of being an astronomer among astronomers. Such a society was a requirement if the transmission of information was to be enhanced. In the Royal Society, that necessary feedback was missing. Just as astronomy began to prosper in Canada, a new, vital organization made its appearance, the body to become known as the Royal Astronomical Society of Canada (RASC).

Amateur astronomers and others interested in the science had congregated in a variety of societies throughout the century, but no strictly astronomical organization was founded until 1868, the year of the first meeting of the Toronto Astronomical Club.[58] On 1 December, eight men met at the Mechanics' Institute and decided to meet monthly to discuss advances in astronomy, to observe, and to give papers on astronomical subjects. A number of the original members would try again after the group's early demise. Andrew Elvins (1825–1918), a Cornish-born tailor, was the moving figure behind the club, which renamed itself a society in May 1869. Self-taught and modest, he was able to bring together interested parties, not only in 1868 but later. Other early members included Mungo Turnbull, a Scottish cabinet-maker; Daniel K. Winder, a one-time lecturer in astronomy who had gone to Toronto from Cleveland as a Civil War draft resister; James L. Hughes, a teacher at the Toronto Model School; Samuel Clare, a teacher at the Normal School; Robert Ridgeway, a teacher at Jarvis Collegiate Institute; George Brunt, an accountant; and Charles Potter, an optician like his brother, Augustus F. Potter.

Evidently, the Toronto Astronomical Society – it was known by other names as well – grew little beyond the original eight members and met irregularly for several years. None of the original members was drawn from the University of Toronto. Whether this was due to lack of public relations by the members or whether other societies, particularly the Canadian Institute and the Literary and Scientific Society of University College, were strong competition is difficult to say. In 1879, the group adopted the name of the Recreative Science Club, and its meetings were somewhat better attended; Elvins and Turnbull were the only founding members still active. About 1882, the notable amateur A.F. Miller joined the group, followed, about 1890, by George F. Lumsden, assistant provincial secretary of Ontario. This led to a decisive turn of events. Lumsden counselled incorporation, which was finalized in March 1890.

The new group, the Astronomical and Physical Society of Toronto, was larger, better organized, and, despite its name, national and international in scope. It had corresponding members in various parts of Canada and, though it may have seemed presumptuous, asked a number of astronomers to be honorary members, including Astronomer Royal W.H.M. Christie, British astronomers William Huggins and Joseph Morrison, the Canadian Sandford Fleming, and the Americans Daniel Kirkwood, Simon Newcomb, and E.S. Holden. A connection with the Toronto Observatory was forged by naming Charles Carpmael the first president and by electing his assistant, Frederick Stupart. Carpmael reciprocated by allowing the members to view with the observatory's Cooke telescope and equipment. By the mid-1890s, attendance at meetings reached upwards of 150. A close connection with the University of Toronto was made in 1892 with the election of Clarence Augustus Chant. He was to become the society's president (1904–7) and founder-editor of its *Handbook* and *Journal*.

From 1890, the society published annual *Transactions*, under the editorship of Arthur Harvey. These were superseded in 1907 by the *Journal of the Royal Astronomical Society of Canada*. During this formative period, the society, which renamed itself the Toronto Astronomical Society in 1900, had members across Canada. Associated societies arose only in nearby small towns, the first being the Meaford Astronomical Society, founded by Rev D.J. Caswell in 1893. At the same time Dr J.J. Wadsworth organized a branch in Simcoe. Other branches followed in Tavistock (1896), Orillia (1897), Hamilton (1901), and Owen Sound (1903). These associated groups, except for Hamilton, disappeared when the society reincorporated in 1902 as the RASC, after which official, semi-autonomous centres, beginning with Ottawa, were formed.

From the time of its resurrection in 1890, the Astronomical and Physical Society of Toronto behaved like a national organization, akin to an earlier version of the RAS in Britain. It was not just a group of amateurs who came together to exchange notes, but rather a society that from the start forged links with professionals elsewhere. An interesting exchange of correspondence between members of the society, chiefly Lumsden and John Collins, and J.E. Keeler of the Allegheny Observatory in Pittsburgh survives.[59] Keeler was a noted spectroscopist who, for the last two years of his life, was director of the Lick Observatory. As an honorary member of the Toronto society, he was asked many questions that arose at the meetings, and he cheerfully responded. He also dispatched a paper for publication in the *Transactions*.

One subject of great interest to the society was the question of international time, thanks to Sandford Fleming's influence. In 1896, the society sent a circular to the governments of all major powers asking whether they would consider employing one standard day, beginning at 0^h00^m Greenwich Mean Time, the practice to begin on 1 January 1900.[60] This proposal had already been broached at the Washington Conference on Standard Time in 1884 but with no resolution. The astronomical day had traditionally begun at noon, but the Canadians, and Astronomer Royal Christie, sought the union of astronomical, nautical, and civil days. The Toronto circular received a mixture of replies: the Americans were opposed, the Austrians cautious, the British willing, as Greenwich already observed the system, and the Germans silent. The proposal failed, and final adoption of the system, known as Universal Time (UT), did not come into effect until 1925.

The Toronto society is an interesting example of an institution that shows its Canadian character. The normal evolution of scientific societies followed a line beginning with a large, comprehensive society. When the number of members having a particular specialty grows to an extent that they feel compelled to break off, a new specialist society arises. In Britain, the Royal Society of London, founded in 1660, though with antecedents, covered all scientific topics. In 1820, the RAS was formed to cater to that science and comprised both amateurs and professionals. It, like its sister group, the Geological Society, was a prestigious association with a royal warrant and fellowships. In time, however, the professionals dominated the group, and the amateurs felt the need for a separate organization. This came about in the 1890s with the formation of the British Astronomical Association.

In the United States, the path was similar; its comprehensive society was the American Association for the Advancement of Science (AAAS), founded just before mid-century. By 1899, the first meeting of the Astronomical and Astrophysical Society of America was held. It was strictly for professionals; a national organization for amateurs followed later.

The Canadian situation differed. There was no comprehensive society – indeed, there still is nothing akin to the AAAS – with a national character,[61] although there was, in Toronto, the Canadian Institute. The original Toronto Astronomical Club was, in a sense, an offshoot of that group. Many of the members of the reconstituted Astronomical and Physical Society at Toronto in 1890 were also members of the Canadian Institute, while Carpmael, Harvey, and Stupart were also fellows of the honorific Royal Society of Canada. With its preponderance of amateurs, the Toronto group was not like the RAS or its American counterpart and was not, with its professional members, much like the British Astronomical Association. It was a unique creation, local with national pretensions, fitting the needs of a country with only a few amateurs and professionals.

Further, the publications of the society were unique. The *Transactions*, later the *Journal*, was not a fully professional journal like the *Monthly Notices* or the *Astronomical Journal*, but neither was it an amateur bulletin. It was a bit of both and has remained so to the present. The ideology of the society, that amateurs and professionals can interact positively within one institutional framework, has worked for nearly a century. If that ideology is realizable and useful, then the smallness of the astronomical interest group in Canada has worked in a beneficial way.

Science education was slow to develop in Canada, partly because of the country's small population, but also because a great deal of time was spent arriving at a consensus on the nature and goals of university-level education. Before the halfway mark of the nineteenth century, a struggle between the upper class and the middle class over the control of university curriculum was waged in most provinces. The original Loyalist and upper-class British immigrants naturally wished to retain their traditional form of higher education, based upon the classics, represented in contemporary Oxford and Cambridge. The Anglican colleges, such as the King's colleges in Windsor, Fredericton, and Toronto, were created by educators and divines as if the Church of England were the established church in Canada, and their curricula reflected the older models of England and the American colonies. Science had little or no part in this model of higher education, which was oriented towards the production of clergymen, professionals, and, above all, gentlemen.

Ranged against this powerful minority view were the more populist, democratic views of American farmer immigrants, merchants, and non-Anglican Protestants, primarily Presbyterians, Methodists, and Baptists. They took their cues from the United States and Scotland, where practical education was dominant, if not always in practice, certainly in spirit. Science, seen as a form of knowledge that could lead to economic power, as well as a means to exalt the Protestant view of God and his relation to man, found a natural place in the curricula of schools founded and guided by people of such bent.

Astronomy is a science that happily combines both an element of the practical – in navigation and surveying – and those higher intellectual and spiritual elements so highly prized by nineteenth-century educators. Not surprisingly, it is in those schools where the more democratic and pragmatic views held sway that astronomy thrived. While we might have expected more sustained study of astronomy in the universities of Canada's metropolis, Montreal, and its second city, Toronto, both institutions were long dominated by upper-class elements. McGill, as a centre of the arts and sciences, remained practically moribund until the arrival of J.W. Dawson, who personified the anti-Anglican, middle-class ideology that would eventually build the university into a world-renowned centre of science and medicine. At Toronto, Bishop Strachan's legacy of a little Oxford in a provincial town lasted long after his death and the legislature's moves to secularize the university. Not until late in the 1880s did a new generation begin to shift the university towards the North American style of higher education. The smaller universities and colleges could never cultivate the sciences, particularly astronomy, to any extent because of lack of resources. Most of the small schools, too, were located in minor urban centres where intellectual sustenance was always difficult to maintain.

Among francophones in Quebec and other parts of Canada, the educational system diverged sharply from the traditions of anglophone Canada. The colleges, following French practice, included astronomy in the curriculum, but the sciences were never intended to be part of a practical education in the sense that a Methodist or a Presbyterian would understand the concept. The Roman Catholic hierarchy was not opposed to the study of science – in fact, Canada's first cardinal taught astronomy at the Séminaire de Québec – but it failed to grasp the value of science in the industrialization of nineteenth-century Canada. Thus, science education fell into paralysis at mid-century and did not emerge until well into the present century. That attitude, coupled with the long tradition of young men entering the liberal professions, ensured that French Canada did not supply anywhere near its quota of scientists to the roster of nineteenth-century Canadian scientists.

Canadian higher education during the nineteenth century was similar to American practice, although its relative size and resources meant that far fewer scientists were graduated. The PH D came later to Canada, and the few Canadians interested in pursuing astronomy beyond the first degree had to resort to foreign schools. In fact, thanks to the nature of universities and government positions, most Canadians who became astronomers did not require graduate studies. As we have seen, virtually all of Canadian astronomy of the last century was practically oriented, thus requiring men with educations sufficient for practical purposes; further training in theoretical branches of the science was redundant. Canadian schools, which produced the majority of Canadian astronomers, were well suited for the task of forming them intellectually.

Amateur and popular interest in the science also reflected Canada's colonial nature. There are few traces of amateur astronomers: few people had the leisure for amateur studies, instruments and books were harder to obtain and more expensive than in England, and, I suspect, the harsh winters were not conducive to star-gazing. The intellectual climate of Canadian cities and towns was reminiscent of the British provinces, where upper and middle classes came together in societies to create what activity they could muster, being so far from the great centres. These activities, especially the reading of papers before such gatherings, speak of a strong yearning for better society than was easily found in such a relatively backward colony. Indeed, we find similar activities in the American midwest and west, themselves cut off to a large extent from the social and intellectual centres of the east. Maturity was long in coming: the precursor to the RASC did not become a permanent organization until 1890 and even then combined both professional and amateur interests under one societal umbrella.

Before the turn of the century, although some cities like Montreal and Toronto had most of the trappings of contemporary centres in Europe and the United States, the country was still small, raw, young, and developing. Astronomy was cultivated little beyond its practical advantages. That any of the more intellectual aspects of the science were kept alive in Canada is due to those who wished to rise above the colonial context and to strive to re-create the culture of more developed nations.

PART THREE
Central Institutions
1905 – 1945

Dominion Observatory (photo by W.J. Topley), c. 1912

A portable transit instrument of the DO, c. 1910

DO eclipse expedition to North West River, Labrador, 1905:
C.A. Chant and J.S. Plaskett (sitting, left and centre in front),
W.F. King (standing, centre)

Opening ceremony, Dominion Astrophysical Observatory, 1917;
J.S. Plaskett in foreground

David Dunlap Observatory, 3 June 1938

DDO 74-inch telescope in Grubb-Parsons factory, Newcastle-on-Tyne

Coelostat house, DO, 1938; spectrograph was in main building.

Ralph E. De Lury adjusts coelostat mirror, DO, 1943.

R. Meldrum Stewart, dominion astronomer, at his desk, 1943.
The dome containing the photographic refractors can be seen
through the window.

CHAPTER FOUR

The Dominion Observatory

We can speak confidently of Canadian astronomy after 1905 as a distinctive and unique entity, both socially and scientifically. The leaders of astronomy for the next two decades – King, Klotz, Plaskett, Stewart, and Chant – were all Canadians, and all of them were proud of the efforts they and their countrymen were making. To be sure, one still finds the rhetoric of 'imperial science,' or the great international community of scientists – the successor to the Renaissance republic of letters – for those were, and are, powerful ideological concepts. The world's astronomical community saw Canadian astronomers (as opposed to British astronomers in Canada) whom they considered their equals.

With the gradual replacement of the old network of observatories under the control of the Canadian Meteorological Service by the Department of the Interior group under W.F. King, we enter a different phase of Canadian astronomy. This new era is dominated by three central institutions: two within the federal government – the Dominion Observatory (DO), in Ottawa, and the Dominion Astrophysical Observatory (DAO), near Victoria – and the University of Toronto's David Dunlap Observatory (DDO), in Richmond Hill. Until the 1960s, most Canadian astronomers worked in one of these three observatories or had some connection with at least one of them. On the whole, the interaction of the three institutions was remarkably harmonious. One rarely comes across evidence of the kinds of dissension that occasionally surfaced in other great observatories and departments of astronomy in the United States and Europe. Indeed, the co-operation among the three – often necessary because of the small size of the professional community – is a strong feature of Canadian astronomy until the 1960s. A number of astronomers moved freely from one institution to another. Not untypically, R.K. Young went full circle from training at Toronto, to a post at the DO, then to the DAO, and finally back to Toronto, where he took part in the planning of, and later directed, the DDO.

This chapter charts the rise and growth of the DO to the 1940s; chapter 5 examines the development of the DAO; and the DDO is described in chatper 6 in the context of university astronomy. All three observatories are so essential to the formation of modern Canadian astronomy that it is convenient to discuss the research work done in Canada within the context of its institutional framework. A few notable astronomers before the 1950s were connected with the universities; their work is discussed in chapter 6. This approach requires taking a certain amount of liberty with chronology, but the roles of the observatories must be emphasized.

Two of these institutions, the DAO and the DDO, were created expressly for astrophysical research, not for practical astronomy. The DO was established for practical astronomy, but through the encouragement of King and the dynamic leadership of Plaskett, it also became a centre for astrophysics, the nucleus for the DAO, and an inspiration for the founding of the DDO. After Plaskett's removal to Victoria, astrophysical research at Ottawa receded, and practical astronomy, which had progressed all along, reasserted itself. Practical work, after the formation of a separate Geodetic Survey of Canada, fell into two major areas: geophysics, and time and positional astronomy, the traditional fare of national institutions. Important as these two areas are, the former will be left to a future historian of Canadian geophysics, and the latter, already chronicled in the work of Malcolm Thomson, will be touched upon only lightly in this work.

While I am not enamoured of 'great man' theories of history, I am struck none the less with the pivotal roles played by J.S. Plaskett and by Clarence Augustus Chant, who established astronomy at the University of Toronto and set up the DDO. Chant has been called the 'father of Canadian astronomy,' but it should become evident that the two men played complementary and equal roles. Chant created modern astronomical education in Canada, and Plaskett, by his leadership and example, created the climate and set the standards for astrophysical research. Any analysis of the growth of the Canadian astronomical research community before the late 1930s must assign central roles to both.

The Dominion Observatory opened officially in April 1905, although research had begun earlier and the staff effectively assembled by 1903. The longitude survey work had continued apace since the 1890s and culminated in the trans-Pacific survey of Otto Klotz and his assistant, F.W.O. Werry, in 1903–4.[1] Canadians had been responsible for a section of the world-girdling 'All Red Cable' (British Empire) route connecting Canada with Australia, the cable having been laid across the Pacific by the end of 1902. When final checks had been made in Australia, Klotz found that difference measured from both east and west of Greenwich amounted to only a fifteenth of a second of time.

Up to this time, longitude work had been undertaken with the telegraph, but in 1902 King learned of the government's intention of working out an agreement with Marconi to test his wireless telegraph equipment for the lighthouse service. In a memorandum to T.G. Rothwell, the acting deputy minister, King argued that the longitude net for Canada could be increased with wireless methods and that two sets of apparatus could be obtained for £474 if Marconi would waive the royalties for scientific work.[2] James Smart, the deputy minister, concurred. However, the minister of finance, W.J. Fielding, upon finding that Marconi felt his directors would be opposed, suggested to King that the Department of the Interior might negotiate with Marconi separately.[3] This was evidently not pursued, and wireless telegraphy for longitude work did not finally become part of the DO's program until 1920. This episode is an indication of the far-sighted leadership provided by King.

The far-sightedness shows up again with the encouragement of J.S. Plaskett. Hired in 1903 as a mechanician, to be assigned the care of the instruments of the observatory, he was named 'astronomer' in the 1905 initial organization of staff. Only Klotz had previously held this rank. Until 1904, all staff except King and Klotz had been temporary, on annual contract. As of 1 July 1904, twelve permanent positions were created, and a budget of $9,299.73 was expended for 1904–5. Before actual research divisions had been established, activities fell into two areas: the time service, with R.M. Stewart in charge, and longitude work, being undertaken by Klotz, Werry, and F.A. McDiarmid. Boundary surveying by J. McArthur and C.A. Bigger constituted a separate branch. J. Macara, the chief computer, undertook reductions. The remainder of the staff included Louis Gauthier, the records keeper, J.D. Wallis, photographer, and W.M. Tobey, observer, and an office staff comprising Wilbert Simpson, secretary, and J.H. Labbé, correspondence clerk.

The first astrophysical project of the new observatory was assigned to Plaskett. This was the impending total solar eclipse of 29 August 1905, which would be visible in Labrador. The expedition came about owing to the initiative of the RASC, which passed a resolution urging the government to act. This was forwarded to Prime Minister Laurier by the society's secretary, J.R. Collins, on 19 November 1904. The society requested the privilege of sending its own observers at government expense, 'the extension of such a privilege to a national astronomical society being entirely in accord with the custom which has obtained in all previous eclipse expeditions dispatched by Great Britain.'[4] Laurier readily agreed. King sent a memorandum to Sifton in December. outlining the observatory's plans and projected a cost of $750 for the expedition.[5] On 29 December the Privy Council approved the costs.

As the only solar equipment in possession of the observatory was a solar camera for the 15-inch refractor, Plaskett had to obtain portable solar cameras and a coelostat – a flat mirror for tracking the sun and reflecting its image.[6] The latter, 20.5 inches in diameter, was furnished by Brashear. A 45-foot-focal-length camera with a 5-inch objective was constructed by Grubb; a 10-foot camera with twin 4-inch Grubb objectives and an 82-inch camera with 4.5-inch Cooke objective completed the cameras. For spectroscopic observations, Plaskett was to take the three-prism slit spectrograph of the observatory for coronal observations and a concave grating spectrograph and prismatic camera for reversing-layer photography. He paid particular attention to the photographic plates, since he had previously experimented with a variety of colour emulsions, both at the University of Toronto and at the observatory.

The expedition, led by King and including Plaskett, Macara, Gauthier, and W.P. Near of the observatory, was joined by a group from the Greenwich Observatory led by Walter Maunder. The RASC group included Chant, Collins, G.P. Jenkins, Prof A.T. De Lury, J.E. Maybee, and Reverend Dr Marsh.[7] Several others, including Prof Louis B. Stewart of Toronto, accompanied the group. After all the work, the site at the Northwest River was overcast on the day of totality. None the less, the expedition gave Plaskett valuable experience with the new equipment, which could then be applied to the solar and stellar programs in Ottawa.

The work that made the observatory world-famous was astrophysical, but it must be remembered that the raison d'être for establishing the institution was utilitarian. Practical work fell into three broad divisions: time service, surveys, and geophysics.[8] The time service, headed by Stewart, worked steadily at improving the timekeepers at the observatory and expanding the network of clock dials in government buildings in Ottawa. This service was initially limited to Ottawa because time services in other parts of the country were entrusted to the Meteorological Service observatories, primarily McGill. In fact, not until 1941 was DO time considered official time for the entire nation. Essential to a time service is the constant observation of star transits, the meridian crossings of specified clock stars. The 6-inch Troughton and Simms meridian circle which had been ordered was delivered in 1907 but, because of defects and other problems, did not go into service until 1910. In anticipation of the work, Stewart had secured the appointments of D.B. Nugent in 1907 and Charles C. Smith (1872–1940) in 1908. The latter succeeded Stewart as superintendent in 1923, with Nugent eventually filling the position upon Smith's retirement in 1937.

Apart from the transits for time observations, the time service, in 1911, undertook a program to determine the right ascensions and declinations of some 3,162 stars, the positions of which could be employed in field longitude and latitude work. By 1923, and some 28,000 observations later, this program was completed. The observatory agreed to participate in an International Astronomical Union (IAU) project launched in 1922 and aimed at obtaining positions for a list of stars compiled in 1914. Ottawa chose 1,368, the so-called Backlund-Hough stars. This task was not completed until 1950, though largely superseded with the publication of Boss's *General Catalogue* in 1937.

The observatory was initially equipped with four pendulum clocks for the time room in the basement, two sidereal clocks by Riefler – which was the standard – and Howard, and two mean time clocks by Riefler and Borrel. Until the meridian circle became operational, star transits for time were taken with the observatory's Cooke transit instrument. Time signals were relayed electrically to city buildings. By 1930, some 700 clocks in Ottawa derived time from the observatory. In 1929, Stewart, then dominion astronomer, obtained a highly accurate Shortt clock as the standard sidereal timekeeper.

Latitude and longitude work continued as in pre-observatory days. Klotz and Stewart had remained in the field until 1905, when they were replaced by other observers. The bulk of the effort was oriented to boundary surveying, as King had been entrusted with this as well as the observatory. Much of this surveying, especially for the Alaska boundary, was undertaken by F.A. McDiarmid, with the assistance of two Toronto graduates, W.C. Jaques and C.A. French. In the summer of 1905, the Department of Militia and Defence approached King to establish a trigonometrical survey of Canada. This was assented to, whereupon C.A. Bigger[9] and J.D. Mc-Lennan were dispatched to begin in the Ottawa region. McLennan died in 1907, but Bigger continued.

In 1909, the government decided to make provision for permanent work, and an order-in-council established the Geodetic Survey of Canada with King as superintendent. Such was King's stature that by 1909 he was chief astronomer, director of the DO, superintendent of the Geodetic Survey, and responsible for boundary surveys. It was the intention of the Department of the Interior that he also be director of the projected DAO. So much power in the hands of one civil servant could not last beyond the lifetime of King; thus, in 1917 the Geodetic Survey was separated from the observatory, with J.J. McArthur, of the boundary survey staff, temporarily in charge. The superintendency passed to Noel J. Ogilvie in October 1917. All geographical work was transferred from the observatory to the Geodetic Survey in 1923.

King's early interest in wireless telegraphy for longitude work has been noted. In 1914, some experiments were made, and in 1921 field tests were performed in the Northwest Territories. Thereafter, wireless became a permanent feature of the observatory, culminating in the CHU radio time signals. However, until that time, field-work was carried on as before with portable transit instruments and telegraphed time signals. By the time of Klotz's death in 1923, the research that he and King had begun nearly forty years earlier, which had led to the establishment of the Dominion Observatory, had passed out of the hands of the national observatory.

The third area of scientific work, that of geophysics, was Otto Klotz's province. He had taken an interest in gravity measurement before the observatory was built, obtaining two half-seconds Mendenhall pendula in 1902. With these he made measurements in Washington, DC, and in Ottawa. In 1905 Klotz continued field-work with Prof Louis B. Stewart of Toronto, but, by 1908, the pendula were diverted to field longitude studies. For several years thereafter, no staff could be released to carry on the effort. Eventually, regular measurements were made and compared with those of Washington, Greenwich, and Potsdam. Klotz also involved the observatory in magnetic surveys: beginning systematically in 1907, field-work was in the hands of George White-Fraser, J.W. Menzies, and, later, C.A. French. During the winter, the magnetic survey staff worked in the Toronto Magnetic Observatory, by then relocated in Agincourt, Ontario. With the aid of summer assistants, the magnetic survey staff had occupied some 1,000 sites by 1930.

Klotz was primarily interested in seismology and had obtained a Bosch seismograph with photographic recording, the only such instrument on the continent. After the devastating San Francisco earthquake of 1906, the AAAS established a seismology committee of which Klotz was a member. Always eager to participate in international co-operative efforts, he induced the government to join the International Seismological Association, which was agreed to by an order-in-council in 1907. He attended its first meeting at The Hague in September of that year. The observatory issued bulletins on earthquakes recorded and, later, established seismographs in Halifax, Saskatoon, Shawinigan Falls, and Seven Falls.

That practical astronomy should have flourished at the DO is no surprise, but the success of the astrophysical side was, perhaps, somewhat unexpected. King had fully intended that some astrophysical research should be undertaken, and it was a stroke of fortune to hire J.S. Plaskett, who, when he joined the staff, had a reputation only in mechanics and photographic studies.[10] The original astrophysical equipment was modest but adequate for restricted work. Brashear had supplied a wedge photometer for measuring stellar brightness, but Plaskett found it inadequate and had a new one built in the machine shop in 1907. Since the observation of binary stars was then the chief work of large refractors, especially at Yerkes and Lick observatories, a position micrometer had been obtained from Warner and Swasey. Although the limitations of this effort were great, Plaskett detailed it to Robert M. Motherwell (1882–1940), recent graduate of Toronto and student of Chant, when he was appointed in 1907. Motherwell was to observe those double stars with few published measurements and whose components moved rapidly enough to require frequent observation. King suggested that occultations be observed, as well, and Motherwell was given this task to perform with the observatory's 4.5-inch Cooke equatorial.

The original equipment had included stellar and solar cameras for the 15-inch telescope by Brashear, but during the first year neither was employed. Plaskett's interest was in spectroscopy, particularly in spectroscopic binaries, emerging at that time as an important topic for professional astronomers such as W.W. Campbell at the Lick Observatory. The original instrument was a three-prism universal spectrograph similar in design to those in use at the Lick and Allegheny observatories. The three prisms could be removed and a single prism or Rowland plane grating substituted. Plaskett found this instrument unsuitable and embarked on a series of experiments over the next few years that made him one of the world's leading experts in spectrograph design. To get an idea of the instrumentation and observational programs in other institutions, he obtained permission from King to make a tour of American observatories in 1906. Lick was his first and most important visit. While there, Plaskett saw the latest changes to the Mills spectrograph of the 36-inch telescope. Some of these changes appeared later in Plaskett's instruments. This trip also included George E. Hale's solar observatory at Mt Wilson, the Lowell Observatory, and the Yerkes, Allegheny, Flower, Harvard, and US Naval observatories, as well as US government departments.[11] During the spring and summer of 1906, Plaskett began designing a new spectrograph to which he applied some of the knowledge gained that summer; he noted, however, that 'the groundwork of the design and many of the details are new and were developed from a consideration of the requirements, and from a knowledge, founded on experience, of the essential features in the designs of a spectrograph for radial velocity work.'[12]

Plaskett had already been in touch with Campbell concerning various problems and had requested suggestions as to a research program.[13] The new spectrograph went into service in 1907. It had several novel features: the slit jaws were made knife-edged – they were typically not at that time – so as not to displace spectral lines; the camera lens, designed by C.S. Hastings of Yale and made by Brashear, employed two light crown elements. Chromatic aberration was overcome by tilting the plates. To overcome flexure and rigidity problems, the frame was built of tubes in a triangular frame similar to Brashear's instrument but coming together much more closely. By making changes in the optical train, Plaskett was able to reduce exposure time by 30 per cent with one prism, which, employed on an instrument that could not effectively work at much less than fifth magnitude, was a significant advance.

During 1906 Plaskett was joined by William E. Harper (1878–1940), a graduate of Toronto and RASC gold medallist there for highest standing in honours astronomy. Harper collaborated with Plaskett in radial velocity – motion in the line of sight – studies on spectroscopic binary stars. At that time some 150 spectroscopic binaries had been discovered, but only about twenty orbits had been calculated. Some two-thirds of those known were young, hot stars of early spectral type. These were to occupy both Plaskett and Harper for many years at Victoria once they possessed an instrument capable of obtaining fainter spectra. Their first DO project was to calculate the orbit of α Draconis, completed in July 1907. By that time they had obtained some 600 plates of 12 stars. Harper, in particular, relished this line of work. Thanks to the new spectrograph, about half of the plates could be taken of early-type stars, for, although the dispersion was about three-fifths that of the Brashear instrument, the exposure time was about half, with more lines visible.

With the radial velocity studies under way, and Motherwell busy on micrometry and occultations, Plaskett brought in W.M. Tobey, one of the staff's original observers. Since Plaskett had dispensed with the Brashear wedge photometer, a Zollner-type photometer, using a Nicol prism instead of a wedge, had been built in the machine shop to his specifications. He had written to E.C. Pickering, director of the Harvard College Observatory, a noted centre for stellar photometry, for suggestions as to what to observe.[14] Tobey made a start on long-period variable stars in 1907–8, but the effort was soon discontinued.

Astrophotography with the original Brashear 8-inch doublet with 40-

inch focal length was undertaken at intervals by Motherwell. As the camera had to be attached to the 15-inch refractor, any photography was at the expense of spectrography. In 1909 Plaskett inquired of Pickering whether an objective prism might be mounted on the camera.[15] Before anything could be done, Motherwell discovered that the lens had aberrations, and the lens was refigured in time for photography of Halley's Comet in 1910. The staff decided to erect a separate building for astronomical photography. The 8-inch doublet was returned to Brashear and was mounted, together with a 6-inch doublet, a 3.3-inch Zeiss-Tessar lens, and a 4.5-inch guide scope, and delivered in 1912. Unfortunately, the small building with a 14-foot dome just south of the main building was not completed until 1914.[16]

Motherwell's work was never more than a peripheral activity of the astrophysics group, which focused on radial velocities and binary star orbits. Plaskett had been singularly successful in his dealings with King. In addition to permanent positions for Harper and Ralph De Lury, he was able to obtain temporary assistance for the radial velocity work, particularly in measuring and reducing observations for Harper and himself. Most of the research in 1907–8 was undertaken with Plaskett's new spectrograph, and the plates were measured on a Toepfer measuring microscope. Early in 1908 the observatory obtained a Hartmann-Zeiss spectro-comparator for solar-type stars. Some of the measuring was detailed to C.R. Westland, and the reductions to F.W.O. Werry, who, having contracted tuberculosis, had come in from the field.[17]

At this time Plaskett's investigations into slit width began. Over the next few years a number of papers were published in the *Journal* of the RASC and the *Astrophysical Journal* and read before the American Astronomical and Astrophysical Society on the probable errors in spectrographic work. Plaskett continued his investigations on various camera lenses and found that one could open the jaws of the slit of spectrograph up to 0.05 mm instead of the customary 0.025–0.031 mm if the jaws were knife-edged; this reduced the exposure time. This research, noted widely, was completed by 1910.[18] Most of the spectrographic work was performed with Plaskett's spectrograph in one of four possible configurations: single prism with a camera focus of 525 mm, or with three prisms with camera lenses of 525, 300, and 260 mm foci. In 1908 Plaskett began to develop another instrument, a single-prism spectrograph with a longer-focus collimator and camera lens for dispersion higher than any of the configurations in the older instrument allowed. Plaskett was able to obtain two more assistants in 1910: John B. Cannon (1879–1940) and T.H. Parker, both recent Toronto graduates and students of Chant. Both were directed to radial velocity.

In 1913, Reynold K. Young (1886–1977) joined the group. He was another of Chant's pupils and an RASC gold medallist in 1909. Chant urged him to go to Lick for further training. Winning a fellowship, Young completed his PH D in 1912, married the daughter of Lick astronomer R.G. Aitken, and then took a position at the University of Kansas. This post was uncongenial, and he later obtained employment at the DO. He remained there until 1917, when he accompanied Plaskett to Victoria. Even before Young's appearance, the radial velocity study was nearing its end. By 1910, some 3,360 spectra had been obtained, 14 oribits had either been, or were in the process of being, calculated. Even so, the new single-prism instrument allowed for only two magnitudes fainter than previously. From this realization stemmed the campaign to obtain a large reflecting telescope for Canada, which was ultimately realized in the creation of the DAO: the account of this campaign will be given below.

When the DO was founded, King and Plaskett intended to embark upon solar research and had planned a solar telescope. The abortive Labrador eclipse expedition of 1905 was a dry run for solar observing techniques. The acknowledged leader in solar physics in North America was George Ellery Hale (1868–1938), founder of the Yerkes and Mt Wilson observatories. Plaskett introduced himself to Hale in a letter in 1906, describing the plans for the Ottawa solar telescope and asking for an opinion. Although most solar observation in those days was sunspot counting – indeed, the 15-inch was used for solar disc photography from 1906 – Plaskett hoped to pursue other lines. As he told Hale, 'The idea of the Director as to the function of a National Observatory is to do work of a routine character which will be useful and which might not be likely to be continued at a private observatory. Of course, there is no objection, indeed he would be glad to have some originality in the work and I would not necessarily be confined to routine observations.'[19] Hale's reply[20] suggested some changes in the proposed instrument, some of which were carried out, and outlined two avenues of research: systematic observations of sunspot spectra and observations of solar rotation. It is the latter that appealed to Plaskett.[21] Further correspondence between them, with suggestions from Hale's associates C.G. Abbott and G.W. Ritchey, ironed out some of the design problems.

During 1907, Plaskett and Harper continued to make observations of sunspots with Brashear solar camera attached to the 15-inch, which was stopped down to 3 inches. In 1907, Ralph E. De Lury (1881–1956), a Toronto chemistry graduate, joined the staff and was assigned to solar studies. His first task was to take over sunspot photography because the coelostat house was not yet finished. This work continued into 1908, daily when possible, but the plates remained unmeasured. The solar image projected by the refractor was 7.5 inches in diameter; the solar telescope would produce a 9-inch image at a focal length of 23 feet. The coelostat house was completed in 1909.[22] The solar telescope, consisting of a 20-inch coelostat, a 20-inch secondary mirror, and an 18-inch concave mirror of 80-feet focal length, fed the solar image to the Littrow grating spectrograph. The plane grating, made by Rowland, would not produce good plates despite a good deal of experimentation by De Lury and Plaskett during 1909–10. A new grating was ordered from A.A. Michelson.

In 1908, before this work began, Hale, who was organizing a meeting of the International Union for Solar Cooperation to be held in Pasadena in 1910, invited Plaskett to participate. No organization in Canada adhered to the union, which moved Hale to write again early in 1910. He suggested that the RASC join and that Plaskett represent both the RASC and the observatory at Pasadena, adding: 'We are very anxious to have you attend, particularly in view of the fact that you are better equipped than any other observatory (with one or two exceptions) to take part in the work of the Union.'[23] This meeting, and a meeting of the American Astronomical and Astrophysical Society in Cambridge, Mass, two weeks earlier, propelled Plaskett and the observatory into the front rank of North American astronomy. At the Harvard meeting, W.W. Campbell proposed that a committee for co-operation in radial velocity studies be formed. Plaskett, who had made a good impression with a paper on the probable errors of radial velocity determinations, was elected to the committee, along with other noted spectroscopists – Campbell, E.G. Frost, Frank Schlesinger, Karl Schwarzschild, and H.F. Newall. This committee met at Mt Wilson during the solar conference. On the train trip westwards, Plaskett stopped in Chicago to confer with A.A. Michelson about the new but defective Ottawa grating which had recently arrived; Michelson promised to replace it.

When the radial velocity committee met,[24] Campbell reported that the radial velocities of stars fainter than fifth magnitude were needed and that Lick Observatory would begin on stars between fifth and sixth magnitudes. Most of the members claimed that they had neither the time nor sufficiently powerful equipment to undertake the work. Walter S. Adams of Mt Wilson, where Ritchey's 60-inch reflector had been placed in operation in 1908, was willing to observe the fainter stars. Plaskett, whose telescope could not yield results with stars so faint, proposed to pursue his research on the reduction of light loss in spectrographs. This participation must have been exhilarating for Plaskett, because, as Campbell wrote Adams, in 1911 there were only eight observatories in the northern hemisphere doing radial velocity work, the DO being one of them.[25]

The Solar Union, as the forerunner of the IAU, encompassed subject-matter beyond solar astronomy. One committee struck at Pasadena, to which Plaskett was named, was on the classification of stellar spectra. This committee had the same personnel as the radial velocity committee, with the additions of H.N. Russell, E.C. Pickering, J.C. Kapteyn, J. Hartmann, W.S. Adams, and G.E. Hale. That Plaskett was associated with the giants of stellar astrophysics could not go unremarked. As he pointed out to King in his annual report for 1911: '[My appointment] on three committees dealing with far reaching international astronomical questions is an evidence I take it of the standing our Observatory has already attained by its work. I take it as a personal compliment, as well as a recognition of our scientific standing, to be associated with such men ... in the discussion of and cooperation in the three important and far-reaching problems above dealt with.'[26]

This second committee, charged with finding a common classificatory system for stellar spectra, was nearly unanimous in its acceptance of the Henry Draper system of Harvard, though members each had particular problems with the scheme. Schlesinger, the secretary of the group, solicited further suggestions in November 1910. Plaskett's response of 26 January 1911 supported the consensus but suggested that the red end of the spectrum needed more research before any scheme of subdivision could be agreed upon. He was willing to undertake some of this work.[27]

The third committee to which Plaskett was appointed was more closely connected with ongoing research at Ottawa. This was the committee on solar rotation. Adams had recently completed a study of solar rotation but wanted a more systematic undertaking. The committee decided to divide the solar spectrum into sections for different observers and that the radial velocities be observed at 15° intervals of solar latitude. In addition, each observer was also to make measurements in a common region, from 4220Å to 4280Å, the centre of Adams's region of measurement. The sections were apportioned to those willing to assist, including the observatories of Pulkova, Allegheny, Cambridge, Mt Wilson, Edinburgh, and Ottawa. Plaskett's section was to be in the region of 5500–5700Å.

In 1910, Plaskett and De Lury had taken a number of rotation plates and effected several mechanical changes to the spectrograph. But, as Plaskett wrote to Adams just before the solar conference, the Ottawa results were giving slightly lower values than those of Adams.[28] Another grating from Michelson arrived in December 1910 but was found less useful than the one already in use. In all, some 130 plates were taken that year. De Lury undertook to discover why the Ottawa velocities were less than those of Mt Wilson. He found, among other things, that there were systematic errors in measuring the positions and displacements of lines between himself and Plaskett. Others in the astrophysics division were brought in to measure a set of plates on the observatory's new Toepfer measuring engine, and all measurements differed. The less the experience in such work, the greater the differences were. An idea that occurred to De Lury was that perhaps the sky spectrum contributed to the rotation effect, giving slightly spurious results. This touched off an acrimonious debate. De Lury worked on the problem for several years; Plaskett doubted that sky spectrum was a problem and assumed that personal equation accounted for the differences. He noticed that the Mt Wilson plates measured by Miss Lasby at Pasadena differed from his own, but not systematically. He asked Adams to check it; [29] Adams replied that his own measurements were close to Plaskett's values, but that Miss Lasby's were on the high side.[30]

In 1913, Plaskett's son, Harry Hemley Plaskett (1893–1980), a student at Toronto, began assisting in the solar rotation work. His measurements, too, differed from those of De Lury,[31] but he was convinced that personal equation was the culprit. He sent a circular on the psychological factors in measuring to members of the solar rotation committee in 1914, suggesting some ways in which the members could co-operate in discovering the source of error and ways to correct it.[32] De Lury wrote to Hale asking for advice, the latter suggesting the use of an improved version of Koch's registering microphotometer.[33] In an article in the *Journal* of the RASC in September–October 1914, Harry Plaskett argued that perhaps the solar rotation varied, his father's measuring habits had undergone some change, there might be systematic errors in the plates themselves. He concluded that the personal equation problem ought to be resolved before assuming solar rotational variation.[34] Adams, who was sounded out by the elder Plaskett, cautioned that the war, which the Americans had just entered, would not allow for immediate research on personal equation, though he, himself, doubted this to be the source of error, but thought it rather instrumental differences.[35] By February 1915, Plaskett had remeasured the 1911–12 plates as well as the 1913 plates. The values now agreed with those of Schlesinger but seemed to indicate that personal equation had indeed changed over a three-year interval.[36]

De Lury returned to the problem with an article in the *Journal* of the RASC in 1916, claiming that atmospheric haze caused the differences in rotation values. Charles St. John and Adams rebutted this in the *Journal* three months later; observations made at Mt Wilson did not show any significant differences due to atmospheric scatter.[37] De Lury responded that it was significant, giving up to 4 per cent difference. This was debated at a meeting of the Ottawa Centre of the RASC in December 1916. Plaskett, who was present, argued that the difference was only about 1 per cent at best, agreeing with Adams. Plaskett tended to be somewhat sycophantic when dealing with Adams and hurriedly wrote him disclaiming any part in De Lury's theory, regretting the publication of the paper: 'He is a nice fellow in many ways but very unsatisfactory for true scientific work as he allows himself to become obsessed with theories such as this and neglects to carefully and fully examine them.'[38]

The last blasts against De Lury came from the younger Plaskett, who, after service with the Canadian artillery overseas, was spending a year working with Alfred Fowler at Imperial College in London. He had already criticized the theory in the *Astrophysical Journal* in 1916 and contributed another paper in 1918. In the *Journal* of the RASC in 1919, he reviewed the question, concluding that haze could not cause variation but that perhaps it was due to local changes or currents on the solar surface. De Lury answered this attack, still holding to his views,[39] but the combined attacks of the two Plasketts and Adams must have taken their toll. It may be significant that, although De Lury continued to make some solar obser-vations, his research turned sharply to the study of earth-sun relationships such as the effect of sunspots on terrestrial weather. In fact, three of four papers read by him at the 1919 meeting of the American Astronomical Society at Ann Arbor dealt with these problems. His desultory scientific output over the next decades dealt almost exclusively with such issues. The combination of J.S. Plaskett's removal to Victoria and his self-fulfilling prophecy of De Lury's work combined to effectively end the Dominion Observatory's importance in solar astronomy.

From 1910, the history of the DAO begins. Its early development was intertwined with that of its parent institution. When Plaskett attended the Pasadena meeting of the Solar Union, he was able to see the 60-inch reflector and hear about progress on the 100-inch instrument. His eagerness to participate in the observation of radial velocities of fainter stars, which was quite impossible with the modest Ottawa telescope, led him to think of Canada building a great reflecting telescope. He broached the subject to King on his return, and the latter was enthusiastically supportive. Plas-kett announced his desire in the *Report of the Chief Astronomer* for 1911:

It seems to me to be an opportunity for enhancing our country's reputation that should not be missed, for a telescope, larger than any in use and one which will enable correspondingly fainter stars to be observed, can be obtained at a compar-atively moderate outlay ... With our experience and record in obtaining accurate radial velocities with the smallest telescope in use in this work, there should be no difficulty in making, with the largest instrument, an unrivalled and exceedingly valuable series of observations; and also, for Canada's Observatory, a reputation second to none.[40]

There was nothing modest in the proposal, but to create another major observatory within a decade of the DO's founding would require determined persuasion of the government. Plaskett had already alerted Campbell in November 1910 that he had been talking to King.[41] When the Astronomical and Astrophysical Society of America met in Ottawa in 1911, it passed a resolution, written by Plaskett, commending the work already done at the DO and supporting the building of a larger telescope. Some headway had been made with the minister, Frank Oliver, and with officials of the Department of the Interior, but in the October 1911 election the Liberal government of Wilfrid Laurier was defeated. The project would have to be reinitiated.

Plaskett and King began the persuasion of the new minister and garnered support where they could. In May 1912, Plaskett wrote to E.C. Pickering at Harvard asking him to send an official letter to King in support of the project; this Pickering was willing to do.[42] He wrote to Campbell in the same vein, telling him that the Royal Society of Canada had passed a supporting resolution and that another letter to King would do no harm; [43] he, too, complied. At the same time, the American Astronomical Society's resolution, King's memorandum on the subject, and Campbell's earlier letter had been given to the new minister, W.J. Roche. Letters solicited from the astronomer royal and other astronomers went to Prime Minister Borden, and the RASC added a resolution in January 1913.[44] The effort succeeded quickly, and Roche agreed in the spring of 1913 that a telescope with at least a 60-inch aperture could be obtained: Parliament voted $10,000 for preliminary costs and authorized a further expenditure of $50,000. King was able to announce this in the May–June number of the RASC's *Journal*.[45] Plaskett had already drawn up the specifications for a 72-inch reflector and contacted Warner and Swasey and Brashear regarding contracts. These were approved by the minister in October and accepted by the Privy Council on 16 October.[46]

Although Plaskett continued to participate in the solar observations and radial velocity work, much of his time was now taken up in organizing the new observatory and overseeing construction of the great telescope. Harper was dispatched to undertake site testing with the 4.5-inch Cooke refractor. A number of sites were checked during the year, including Ottawa.[47] Plaskett and King had asked Campbell and Hale for assistance, and Plaskett went to several American observatories for consultations.[48] Harper's observations seemed to indicate that a site near Victoria was the most suitable. The American astronomers concurred, although Campbell tried to persuade Plaskett to locate the telescope in the southern hemisphere in 'some tropical English colony,' since there were no large instruments in the south.[49] The largest telescope then in operation was Ritchey's 60-inch reflector at Mt Wilson. Plaskett had its blueprints, but the final telescope for Victoria was superior in several ways, owing to his own design improvements and the ideas of E.P. Burrell of Warner and Swasey. The telescope was the first to have self-aligning ball-bearings in both polar and declination axes and to be fully electrically driven and controlled in motion.

King was to have overseen the construction, but the press of work necessitated handing this over officially to Plaskett. During 1914 a number of decisions had to be made. The site on Saanich Hill, north of Victoria, was approved after Plaskett personally asked Sir Richard McBride, premier of British Columbia, for assistance in building a road. This was received, and a provincial grant of approximately $30,000 allowed for the purchase of fifty acres.[50] Construction of the main building and dome, the director's residence, an office building, and houses for observers was to be dealt with by the federal Public Works ministry.[51] The contracts were to be let in 1915, but a number of political considerations almost scuttled the project.

In February 1915, F.H. Shepherd, MP, in whose riding the observatory site lay, wrote to Plaskett asking him to put pressure on the site engineer, who was responsible for site clearing and water-supply, to hire men from the local Conservative Association rather than men from Victoria.[52] As tenders had not yet been called for, Shepherd wrote Robert Rogers, minister of public works, demanding action since the BC government had already committed itself.[53] It appears that Shepherd must have complained about the hirings to the minister; in March, E.L. Horwood, chief architect in Public Works, fired the site engineer, G. Gray Donald, for not respecting local patronage principles.[54] Tenders were soon accepted by the Privy Council for construction. A vote of $75,000 for buildings was agreed to in July, but in December, Prime Minister Borden wrote to Premier McBride and to Roche, the interior minister: 'I do not think it would be possible for us to undertake the construction of these buildings until the conclusion of the war. The requests for public works, many of which are quite urgent and important, come to us almost daily and certainly weekly; under present conditions we have found it necessary to postpone the consideration of such matters. You will understand how difficult the situation would be if we departed from that principle.'[55] McBride quietly passed this along to Plaskett. Although Rogers agreed with Borden, Roche did not.[56] None the less, work went forward.

The telescope itself was not problematic, excepting the near non-delivery of the mirror blank. As only the St Gobain glassworks in France was willing to undertake large discs – the 100-inch disc for Mt Wilson had been cast there in 1908 – the order was placed with it for a 72-inch primary and a 55-inch testing flat. The mirror blank was shipped to Antwerp for shipment in the summer of 1914 and left the continent in July, just one week before war was declared. As it happened, the St Gobain works were destroyed a few weeks later, during the German invasion of France. The testing flat was never shipped, and Brashear, through Schlesinger, attempted to obtain the 60-inch Common disk from Harvard. Pickering was willing to sell it to Brashear, but Plaskett felt it to be too thin.[57] In the end, a testing flat of plywood was constructed. Plaskett was in Cleveland and Pittsburgh several times during 1915–17 to oversee construction. Finally, in March 1918, the last mechanical parts were shipped by rail, the mirror was silvered and mounted, and first light was obtained on 3 May.[58]

Amid the excitement of building the world's largest telescope, personal relations at the DO were clouded by a power struggle between Klotz and Plaskett. When W.F. King died in 1916, no successor was immediately appointed. Wilbert Simpson, secretary of the observatory, though not a scientist, became acting chief astronomer. Both Klotz and Plaskett were equal in rank, but Klotz was senior and obvious successor. Just before King's death, however, Plaskett had asked to be promoted before leaving for Victoria.[59] Klotz and Plaskett disliked one another, and Klotz disliked Simpson, who was close to Plaskett. The latter, fearing that Klotz would be appointed to head both observatories, wrote letters to his American friends, enclosing a circular describing his achievements and asking that they write to Roche. The circular asked each recipient to state what department and who at the observatory was responsible for the observatory's reputation; the relative importance of the work at Ottawa and Victoria; and who was the best-suited Canadian to take charge of the observatories. The circular went out to Schlesinger, Pickering, Hale, Campbell, and Frost, with S.A. Mitchell writing unsolicited.[60] As Plaskett remarked to Hale, he wanted the directorship at Victoria so long as it was independent of Ottawa but would accept the post of chief astronomer, 'to which I have some claim': 'But rather than have to work at Victoria interfered with by an envious or jealous man here [i.e. Klotz], and of this I fear there is danger, I would accept the latter position (provided of course I can get the appointment).'[61]

The astronomers responded, but Roche, whom Plaskett described as 'several grades above the average politician,'[62] was not to be hurried. His reply to Hale stated that he had no intention of allowing the DAO to be independent of the Ottawa institution but that he would say nothing about King's successor as yet.[63] It is something of a puzzle why the appointment was not made, though the common opinion was that Klotz, with an unfortunate German name at the height of the war, would have to be eased in quietly and later. This was quite unfair, since Klotz, an active member of the Canadian Club of Ottawa, was, despite his German ancestry, an ardent and extremely loyal Canadian. Klotz was beside himself over this political manoeuvring by Plaskett. He noted in his diary in April 1917 that Motherwell, De Lury, and Cannon were in a state of low morale since Plaskett was packing up DO instruments for use at Victoria, although they were needed by the Ottawa astrophysics staff. Evidently, Plaskett had gone straight to W.W. Cory, the deputy minister, for authorization to take the instruments.

Worse was to come. Motherwell went to Simpson, the acting chief, about the instruments. As he related to Klotz, 'When he spoke of a list of instruments – they both laughed in his face. The discussion was made personal by Simpson by nasty innuendo, to which Plaskett listened. Finally Simpson said "I have given him authority to take anything he wants"!!! adding "the interview is over"! Motherwell was furious – he said he couldn't understand Plaskett, as for Simpson – he knows him to be capable of anything.'[64] The deputy minister had been primed, as well. Klotz continues,

In this morning's interview Mr Cory said [to Motherwell] that Dr Plaskett had told him that there was to be *no* astrophysical work done hereafter and it was on the strength of this statement that he was appointed Director of the Victoria Observatory – Motherwell nearly collapsed on hearing this falsehood, and Plaskett knows it to be falsehood, for the Director, Dr King intended the work here to go on as before, besides he intended to be director of both observatories, Plaskett superintendent of the western one.[65]

Despite this wrangling, still no decision was taken on the appointment. J.J. McArthur was acting chief astronomer during the summer of 1917. It rankled Klotz that Plaskett's appointment as director at Victoria was assented to by the Privy Council in April.[66] This had been Roche's personal recommendation, with the minister adding that Plaskett would have 'full control of its policy and work.'[67] This series of episodes shows the darker side of Plaskett's ambition. Although only Young went west with Plaskett when the new observatory opened, stellar astrophysics was crippled in Ottawa and eventually died altogether.

In October Klotz was appointed chief astronomer, but the former powers of King were divided between Noel Ogilvie, taking the Geodetic Survey, and McArthur, receiving the Boundary Commission.[68] With Klotz in charge and Plaskett and Young now in Victoria, the observatory settled down to routine work. In 1918 an 'Observatory Club' was created to provide fortnightly seminars for staff members to exchange technical information that could not be discussed at local RASC meetings. Klotz pursued seismological work; magnetic studies continued at Agincourt and in the field by summer students, among whom was J.A. Pearce. Gravity research, undertaken after some years' hiatus by R.J. McDiarmid in 1914, under the auspices of the Geodetic Survey, was given to A.H. Miller. In 1928, the DO compared gravity values with Potsdam, Washington, and Greenwich.[69] In 1930, this field passed to W.G. Hughson, hired that year. The opening of the DAO meant, apparently, that two institutions would undertake astrophysical research in Canada, but the actual effect was almost the opposite. It was never King's nor Plaskett's intention to transfer the entire astrophysical staff from Ottawa to Victoria, but the removal of first Plaskett and Young, then Harper, had a devastating effect upon the output of the DO staff. This is strikingly obvious if we compare some publication statistics before and after the removal.

The *Report of the Chief Astronomer* ceased with the 1911 volumes and was replaced with the *Publications of the Dominion Observatory*. Volumes I and II, complete by 1916, were dominated by the astrophysics group, with seven papers by Harper, nine by Cannon, three by Plaskett, four by Young, and one by Parker, for a total of twenty-four; the remaining seven were by the geophysics and surveying groups. This pattern was repeated in volumes III (1919) and IV (1920), primarily papers by Harper and Young. But with volume V (1922), five papers were by the non-astrophysical staff and five by François Henroteau, one being written with J.P. Henderson. From that time until 1949, excepting two widely spaced papers by De Lury, virtually all the astrophysical output was by Henroteau and his assistants or by the geophysics group.

A search of W.E. Harper's *General Index* of the RASC's publications shows that the DO astrophysical group published, beyond official publications, 103 scientific articles in the *Journal* from 1907 to 1918, but fully two-thirds of these were by the two Plasketts, Harper, and Young. This same group published 48 articles while in Victoria from 1918 to 1931; Harper and H.H. Plaskett came a little later. The rest of the Ottawa astrophysical staff, Cannon, Parker, Motherwell, De Lury, and McDiarmid, accounted for 37 papers before 1919, but only 24 from 1919 to 1931, and 17 of these were by newcomer Henroteau.

To check, one can survey the index of the RAS's *Monthly Notices* from

1911 to 1931. Those of the Victoria staff who had been elected fellows, both Plasketts, Pearce, Petrie, and Beals, contributed 10 papers to the *Monthly Notices*; the Ottawa fellows, De Lury, Henroteau, and McDiarmid, none. Even if we add the non-astrophysical fellows at Ottawa, Klotz, Smith, and Stewart, we still get none. A comparison of the official publications of the DO and DAO also speaks eloquently of the difference. What happened? There are three possible explanations: first, the organizing genius and example of Plaskett were removed from Ottawa. There was no obvious successor of his calibre to offer leadership. Second, the group left behind worked under the directorship of first Klotz and then Stewart, able scientists not interested in astrophysical problems. Third, one could argue that the quality of the men left behind was simply not up to the level of the group that departed. This may have been true in some cases, but De Lury and Cannon's published work seemed to dry up. Cannon had been a promising worker, with 18 papers in the RASC *Journal* before the division, but he may well have been disillusioned by Plaskett's removal. Offered a position in the new Geodetic Survey in 1919, he accepted and left astrophysics altogether. De Lury's change, from 12 papers before 1919 to 3 afterwards, has already been assessed. Apart from Henroteau's work, the DO became an institution for geophysical and time research after Plaskett's departure.

In 1919, François Henroteau, a young Belgian, was appointed assistant astronomer. He was to become the most notable figure in astrophysics for the next few years. His interests were wide and up to date, and he became engaged in the study of long-period variable stars with the 15-inch telescope. In 1920, Stewart and J.P. Henderson made observations of Nova Cygni #3; thereafter, Henderson became Henroteau's chief collaborator. In 1921, Henroteau challenged R.K. Young's work on interstellar gas (see chapter 5). From 1919 to 1921, Henroteau and Henderson observed β Canis Majoris stars, only four of which were known in detail in 1920; one of them, 12 Lacertae, was studied by Young while he was at Ottawa. They also continued the DO's interest in o-type stars, some 24 of which were kept under surveillance. Despite the amateur work on variables by the American Association of Variable Star Observers, and some professional studies, Henroteau believed that a more systematic program was necessary. In 1925, he and H. Grouiller of the Observatoire de Lyon announced a co-operative effort to determine the magnitudes of Cepheid variables by means of comparison stars, which, in turn, were to have their magnitudes accurately measured.[70] He was able to establish the periodicities of a number of β Canis Majoris stars. One curiosity of his work, however, was the inordinately high systematic errors in his radial velocity measurements.

When R.E. Wilson's *General Catalogue of Stellar Radial Velocities*, which employed the Lick Observatory measurements as a standard, appeared in 1953, the corrections to Henroteau's values were − 9.8 km/s, the highest in the catalogue. Struve believes that he might have used a narrow slit for the comparison spectrum and then opened the jaws for the star,[71] though one would not think an experienced observer, who worked with Plaskett-trained men, would commit such an error.

By 1933, only De Lury and Motherwell of the original astrophysics staff were still at the observatory. The former, with the assistance of John O'Conner, continued his studies of the terrestrial-solar link, his chief interest until his retirement in 1946. Otto Klotz, who died in 1923, was succeeded by R.M. Stewart, with the title dominion astronomer. Stewart remained head of the observatory until his retirement in 1946. His position gave him authority over purchases at Victoria, and the exigencies of first the depression, then the Second World War, plus Stewart's own penny-pinching outlook, produced some strain between Ottawa and Victoria.[72] Tight finances were not all Stewart's fault, of course; in 1928, he attempted to have the salaries raised by the Civil Service Commission. He wrote to directors of a number of American observatories for comparative statistics and found that astronomers at Lick already made more than Stewart hoped to get for equivalent-ranked men at the observatory.[73]

The lack of competitive salaries was not the only administrative problem. Stewart needed to fill a post in the astrophysics section in 1929 and received an application from H.F. Balmer, a former student of Chant's who had been at Lick in 1924–6 and was currently at the Goodsell Observatory in Minnesota. In reply to his inquiries, Stewart learned from R.G. Aitken that Balmer was a good observer but not much interested in research, and Aitken offered one of his own graduate students, but Stewart had to reply: 'Our position as a branch of one of the government departments is in some respects a handicap in obtaining the best men for our work, since according to the general regulations it's not possible to appoint anyone other than a Canadian unless we are able to establish the fact that no properly qualified Canadian is available.'[74]

Further changes came in 1937 when the Department of the Interior was dismantled. From 1937 until 1950, both the DO and the DAO were part of the Surveys and Engineering Branch of the Department of Mines and Resources. By the time the Second World War approached, the DO was concentrating almost entirely on the time service, staffed by Smith, Nugent, McClenahan, and Malcolm Thomson. The only astrophysical work that continued was Motherwell's photometry; he had been recruited for the international Cepheid program in 1924.[75] He retired in 1940, dying the same year. The solar program of De Lury and O'Connor continued until the former's retirement. For De Lury, there was some satisfaction after the controversy over his interpretation of the Ottawa velocity differences.[76] By 1922, he was able to discover that part of the cause of the systematic error was oil in the measuring micrometer that caused the nut to move relative to the screw, but he steadfastly insisted upon his view of blended spectra. Finally, in 1938, the Mt Wilson workers admitted that blended spectra did make a difference. However, since 1913, the Ottawa average velocities, taken from measurements of 13 strips of solar image photographed simultaneously, had agreed with those of Mt Wilson.

R.M. Stewart continued as director and dominion astronomer during the war years, taking no holidays, and trying to keep programs going without a full complement of staff and sufficient funds. When he retired, he was replaced, not, as one would expect, by one of the practical astronomy staff but by C.S. Beals from Victoria, who inaugurated a new era at the observatory, which brought to an end the traditional astrophysical work.

At the turn of the twentieth century, Canadian needs in astronomy were being met by the small univesities, the staff of the Canadian Meteorological Service, and the astronomical staff of the Department of the Interior. Astronomical fixings of points for surveyors had settled into a routine; time was provided to several cities and railways by the small observatories, directed from Toronto; and the boundary surveys, along with intercontinental longitude work, were performed by the government astronomers. All of this seemed to be adequate, but a splendid national facility – the Dominion Observatory – opened in 1905 with a staff and instrumentation that exceeded what one might expect of a small, developing country of only five million people. Within five years, the example and energy of J.S. Plaskett propelled the observatory into the 'big leagues' of international astronomy. Within another five, a new observatory, derivative of the first, to be equipped with the world's largest telescope, was under construction in British Columbia. At first glance, this all seems improbable.

Some of the astronomers of the time claimed that the influence of William F. King was primarily responsible for the creation of the DO, but that is unlikely, for although King was certainly a competent astronomer and administrator, he had no widespread reputation outside Canada, because he was, above all, a practical man. He posited no new theories, discovered no new celestial objects, did not champion the rapidly growing field of astrophysics. He was not a political heavyweight in the Liberal party, which had come into power in 1896, and did not have any special influence over successive ministers of the interior or over prime ministers. Wilfrid Laurier, so far as we know, had no especial interest in science, nor was there a strong pro-science lobby in Parliament. Canada's astronomical needs were being met without the addition of a new, and expensive, institution. Why was the DO built?

A permanent base for operations was a natural desideratum of astronomers like King and Klotz, and, during the 1890s, they had tried to influence the department to provide just such a base. Yet something more permanent than the temporary Cliff Street observatory in Ottawa, but more modest than the institution eventually built in the Central Experimental Farm, would have sufficed. The answer lies, I believe, in the ebullient view of Canada's future held by those in power. A long period of financial depression came to an end about the time of Laurier's formation of a government, the wheat boom was just beginning to take shape, and immigration – to swell enormously just after 1900 – was on the rise. It was Laurier who claimed that the twentieth century would belong to Canada, and that optimistic view was echoed, perhaps even more strongly, by energetic leaders from western Canada like Clifford Sifton. The times and the leaders came together at the right moment for William King. Obtaining a first-class institution turned out to be almost easy.

Above all, the Dominion Observatory was a symbol. Parliament, at the behest of its leaders, was willing to provide a third of a million dollars – a vast sum for science in the budgets of those days – because it would give Canada international status. The DO signified national maturity. Ottawa now could boast a national institution for astronomy just like those in London, Paris, Berlin, or Washington. It is probably not coincidental that, after decades of complaints and political manoeuvring, a national museum was also created about the same time. These were visible symbols of nationhood, not just the provincial institutions that graced provincial cities. Canada had recently sent contingents to South Africa; although the Canadians fought under British command, they were part of a Canadian, not British, army. Rising nationalism and industrial and agricultural strength and wealth all contributed to a feeling that Canada was coming into its own. The DO, reflecting support for higher human achievements in science, was a relatively inexpensive way to exhibit this growing maturity and civilization.

In the period up to about 1920, the observatory – and the country – were fortunate to have, among its staff, far-sighted and energetic men like King, Klotz, and Plaskett. King's role is best assessed as supportive: a good administrator, he knew how to set a good example and to support his staff in new areas of research. As the original raison d'être of the observatory was practical, King ensured that Stewart and his co-workers had the best instruments they could obtain to develop a first-rate time service and provide assistance to the surveyors in the field. Klotz was encouraged to move into the growing field of geophysics, to involve the observatory in international conferences and programs. Plaskett, above all, was singled out for the greatest support. There was no prima facie reason for the DO to cultivate the new field of astrophysics; there was no practical payoff for Canada. Yet Plaskett received the additions to staff he desired and support for his international contacts. To a large extent, Plaskett and his co-workers made the reputation of the observatory, but it was King, and the ministry, that supported Plaskett.

The success of the campaign for a new facility, the DAO, reflected not just the lobbying efforts of King, Plaskett, and their allies but also the desire of government and Parliament to show the world that Canada would take its rightful place in the world of science, just as it had done in the war and in the empire. With the passing of King and Klotz, and the removal of Plaskett to Victoria, the Ottawa observatory settled into a routine, providing the practical astronomical services for which it was designed. If its international prominence waned during the years of Stewart's directorship, it was more a reflection of the attraction that astrophysics, Plaskett's forte, held for the astronomical community than a reflection on the quality of the men and work in Ottawa.

CHAPTER FIVE

The Dominion Astrophysical Observatory

The early history of the Dominion Astrophysical Observatory (DAO), near Victoria, provides a striking contrast to that of its parent institution. Where the Dominion Observatory (DO) had seen rapid growth and a building reputation, staff dissensions, and frustrations with instrumental shortcomings, the DAO started in the 1920s with a ready-made reputation in the persons of Plaskett and Young, splendid instrumentation, and comfortable isolation from the Ottawa bureaucracy. Thus, the social history of the DAO can best be described as quiet. The milestones were a steady stream of high-quality research, a slowly growing staff of first-rate astronomers, and quiet diversification of research lines.

Yet, because the observatory undertook only one great branch of astronomy – astrophysics – rather than several separate areas of inquiry, it has been able to maintain its institutional integrity. The DO, after only sixty-five years of existence, was dismantled. After nearly as long a history, the offspring institution is stronger than ever. This turn of events is no reflection on the men who piloted the DO; rather, an institution dedicated to several mutually exclusive areas which overlapped with other departments of government was naturally vulnerable to political and bureaucratic manoeuvring. This was never the case for the Victoria observatory.

From the outset, it was clear that the DAO would not have a large staff. Plaskett was able to persuade only Young to accompany him west, although he wanted Harper as well because all three were deeply involved in radial velocity work. Writing to the latter in May 1918, just after the 72-inch mirror was installed, Plaskett admitted that he could not offer him the assistant directorship, as the staff was too small and the time not right.[1] Harper's reply was that to go west he would need a larger salary – he earned $2,500 at Ottawa – and that if Stewart could be an assistant director in Ottawa, why couldn't he be one at Victoria?[2] None the less, Plaskett tried to pull strings in Ottawa, telling Harper in October that Klotz was resisting a transfer, but

> if you come *now* you will be the senior officer at the observatory, will be left in charge when I am away, and will normally and naturally, practically inevitably, succeed me as director. You have been well treated at Ottawa in the way of promotions and salary increases, better than any other observers, as you must realize, and there was much dissatisfaction among them over it. If you were now promoted to the next class, what could and would Young say whose training and degrees give him at least a talking advantage over you, and who is exceptionally able and industrious.[3]

His persuasion prevailed: Harper arrived early the next year and, as Plaskett had foretold, succeeded to the directorship in 1935.

During 1919, with only three astronomers at the observatory, the observing routine became fixed, and it was followed for many years. Each man had the entire night in the dome and was assisted by the night assistant, at first Tom Hutchinson. The observer took the next day off, and the following night was the second observer's time, and so on. Evidently Hutchinson, although an excellent mechanic, did not work as hard as Plaskett expected, and the latter had to ask Adams at Mt Wilson what that observatory expected of its night staff.[4]

Plaskett planned to continue research along the lines already begun at Ottawa but before the telescope was complete wrote to a number of astronomers asking for opinions of programs that might be undertaken. Young was sent to the Lick Observatory to confer with Campbell in 1917; in the mean time, Plaskett inquired of J.C. Kapteyn in Groningen what he counselled. Kapteyn replied that radial velocities must be catalogued along with proper motions, his own interest, and that these should be divided up by spectral classification. He suggested working through Boss's *Preliminary Catalogue*, which included some 4,500 stars down to sixth magnitude, by sharing the observations among Mt Wilson, Lick, and Victoria for the northern sky and Cordoba and the Cape for the southern.[5] This was, of course, what the radial velocity committee, of which Kapteyn was a member, had discussed in Pasadena in 1910.

Writing to Adams, Plaskett volunteered to undertake the study of some 1,000 to 1,200 of the most northerly stars and related that Campbell, who had recently visited Victoria, was willing to take the brighter stars.[6] Letters to Hale, Adams, and Campbell firmed up the plans. As Plaskett pointed out to Hale, while there were only Young and himself to do all the work at the moment, they were willing to undertake a project of this magnitude because 'this observatory is a national institution and as such should, as I understand it, not be unwilling to undertake routine investigations if carefully planned so as to give the best result in the solution of sidereal problems, even if they do not offer so great opportunities for individual initiative as more restricted investigations.'[7] This was, in fact, the DAO's credo until the arrival of a new generation of astronomers. Plaskett finally decided to work on the Boss catalogue stars north of the celestial equator located on even-numbered minutes in right ascension, plus some stars from the Harvard selected areas.[8]

By April 1919, Plaskett was able to report to Adams that he and Young had exposed 1,800 plates and had already measured 1,000. Late in the year Harper and Harry Plaskett, recently returned from Imperial College, joined in the work. By 1921, a catalogue of 594 radial velocities was published.[3] Hale, on seeing a preprint, remarked: 'It is surprising to see how rapidly and effectively you have conducted this important piece of research, which does immense credit to you and your associates.'[10] This was what Plaskett had worked for, but, inevitably, politics – internal and external – intruded. Young had worked out a table of conversions from right ascension and declination to celestial longitude and latitude to aid in correcting stellar velocities to the sun's motion. This, they hoped, could be printed by the government, but the publications committee in Ottawa, evidently on Klotz's advice, refused to do so. Plaskett fired letters off to Adams and Campbell seeking assistance in convincing W.W. Cory, the deputy minister, to allow publication of the table. Adams readily agreed, but, to Plaskett's surprise, Campbell thought the tables cumbersome and insufficiently accurate.[11] Plaskett eventually convinced him but in the mean time had written Cory directly, complaining:

I am taking this opportunity of expressing my feelings in regard to Dr. Klotz's endeavours to belittle and discredit the work here ... I resent Dr. Klotz being consulted about any of the affairs of this observatory. In the first place he has no personal knowledge of the work or methods as his astronomical training was along entirely different lines; secondly he has no sympathy and finally his actions show he must be jealous of the institution and its equipment and opportunities.[12]

This was the coda to the 1917 difficulties. In any event, the tables did not appear.

Plaskett had succeeded with politicians when the DAO project was first broached, as he always had an eye for a chance. When he learned that the new leader of the Liberal party, William Lyon Mackenzie King, was in Victoria in September 1920, he invited him to visit the observatory.[13] Although King was unable to come, Plaskett tried again in 1923 when he wanted an office building for the observatory. He wrote directly to King, by then prime minister, in March, arguing for the new building but also noting his own recent election to the Royal Society and enclosing a newspaper cutting concerning 'Plaskett's Star,' the most massive star then known. King replied that he was discussing the proposed building with his colleagues.[14] The tactic worked, as $15,000 for the office building was placed in the 1923 supply estimates. With that success behind him, Plaskett approached J.H. King, minister of public works, in 1925 to obtain a new director's house and observers' houses, at the cost of some $30,000. When no action was forthcoming, he wrote directly to the prime minister in January 1927, extolling the work of the observatory and reminding King that, although British Columbia had not voted Liberal in 1926, the observatory was still a federal institution.[15] W.W. Cory was incensed that Plaskett would appeal over his, and his minister's heads, instructing him to deal with the department in future, adding: 'I am sure you will realize that the Prime Minister has plenty of worries of his own as leader of the Government without handling the official business of various Departments.'[16]

During the 1920s, the work branched out somewhat, but the bread and butter remained radial velocities and the computation of the orbits of spectroscopic binaries. By the time volume 1 of the *Publications* was completed in 1922, twenty-six of thirty papers dealt with these topics; the output was enormous, considering that only four astronomers were at work fulltime. Some new lines were explored: Harry Plaskett, more skilled at astrophysical theory than the other three, revised the Harvard classification for o-type stars, dropping the special notations for stars with emission in their spectra. This method was generally adopted.[17]

In the mean time, his father and Young participated in the growing interest in interstellar absorption. In 1904, Johannes Hartmann discovered that the H and K lines of ionized calcium in the spectrum of the binary δ Orionis differed in radial velocity from the stars themselves. In 1909, J.C. Kapteyn predicted that interstellar lines might be observable. V.M. Slipher confirmed this later that year with the Ca II lines, but no one immediately followed up the idea. In 1920, R.K. Young, apparently unaware of Slipher's work, noted that these lines appeared only in early-type stars – specifically in type B3 or earlier – and in spectroscopic binaries. There were three possible interpretations: calcium clouds lie between the star and the observer, Hartmann's view; the calcium cloud is an extended nebulosity; or the cloud is part of the stellar atmosphere above the reversing layer. Young adopted the last view. His objection to Hartmann's theory was that all stars, or at least all B stars, should exhibit sharp H and K lines but do not. The second view was discarded for much the same reason; in addition, Harry Plaskett had found two early B-type stars close together, one with a sharp K line, the other with a diffuse one. One of Young's best examples was 12 Lacertae, a binary he discovered at Ottawa, with calcium lines that oscillated with about half the period of the stellar lines.[18]

Young's paper was disputed by Henroteau at the DO,[19] who argued that clouds of 'metallic gases' lay between the stars; his data, collected by himself and J.P. Henderson, were derived from measurements of the H and K lines of some 42 early-type stars photographed at Ottawa. One of Young's arguments against the interstellar nature of the calcium was that the differences between the radial velocities of the calcium lines and stellar lines were due to measurement error. However, in June 1922, J.S. Plaskett announced in the *Monthly Notices* the discovery of 'Plaskett's Star,' for many years the most massive stellar system known.[20] This binary, HD 47129, with an estimated combined mass of about 140 solar masses, exhibited the sharp H and K lines. The difference in the calcium velocity in comparison with the stellar system was too great to be attributed to measurement errors. He concluded that the calcium could not be part of the system, but relatively at rest in space. This he confirmed with the o-type spectroscopic binary BD 56° 2617 in 1923.[21] In a comprehensive review paper on the o-type stars in 1924,[22] he investigated a number of o- and B-type stars in several regions and was able to show that the velocities of calcium and of sodium lines, noted earlier by Heger at Lick, differed from the stellar lines, sometimes greatly. Discarding Young's and Hartmann's views, Plaskett suggested that the calcium lay in large clouds relatively at rest with respect to the moving stars and that the hot stars ionized the calcium, giving rise to the characteristic sharp lines. This was disputed by Otto Struve, but in 1926 A.S. Eddington provided a theory similar to Plaskett's, which assumed further that the calcium clouds were spread throughout the galaxy. Plaskett would establish this view observationally after his greatest triumph, the confirmation of galactic rotation.

With the publication of radial velocities of the first program, Young and Harper turned to the problem of spectroscopic parallax. This method of determining distance from spectral lines was pioneered by Adams and Kohlschutter at Mt Wilson in 1914. Since the DAO now had a large collection of plates from the radial velocity work, Young and Harper began measuring the lines employed at Mt Wilson, plus a few new ones. This effort culminated in the publication, in 1924, of a catalogue of 1,105 absolute magnitudes and parallaxes.[23] Plaskett, in the mean time, decided to work on the B-type stars.

At this point, however, Young was appointed to the Department of Astronomy at Toronto, and a replacement was required. Plaskett had heard of the work of Joseph Pearce, then studying at Lick, and wrote to R.G. Aitken inquiring whether Pearce would be a suitable replacement.[24] Pearce (b. 1893) had studied at Toronto for two years, then joined the army in 1915, returning to Canada a major in 1919. Taking his BA the following year, he became Chant's assistant during term and worked for the DO and the Meterological Service during summers. At Chant's insistence, he applied for and received a research fellowship at Lick in 1922 and spent two years in California. Aitken hoped that he would finish his PH D first,[25] but Pearce was appointed to replace Young in 1924. Over the next six years, he was fully immersed in the B-type star program with Plaskett. Thus his doctoral dissertation went unfinished. During these years the authorities at the University of California kept providing him with extensions and urging him to complete the requirements and, at one point, suggested that he substitute one of his observatory publications as his dissertation, but Pearce felt this to be unhonourable. Finally, with coaxing from astronomers at Berkeley and from Plaskett, he finished and was awarded his degree in 1930. By that time, he had come to world-wide attention for his collaboration with Plaskett on galactic rotation.

In the mid-1920s, Jan Oort in the Netherlands had become interested in stellar velocities, which seemed to show a strong tendency to move in two different directions. Swedish astronomer Bertil Lindblad had, in 1927, suggested a model of the Milky Way calling for differential rotation, and Oort was quick to find observational evidence that supported this view. He proposed a mathematical expression giving the radial velocity of a star as a function of its distance from the galactic centre and its galactic longitude, locating the centre close to where Harlow Shapley had suggested, by his analysis of the distribution of globular clusters.[26] As there were several competing models of galactic structure and dynamics, confirmation was necessary. This Plaskett and Pearce were able to provide almost immediately.

When Pearce arrived in Victoria, Plaskett divided the new B-star program between them; this was to be a three-year undertaking involving some 1,056 stars of class B brighter than magnitude 7.50 and north of declination −11°.[27] The two shared the work approximately equally, with some assistance from other staff members and summer students, especially W.H. Christie. A catalogue of some 523 O- and B-type stars (B0 to B5) was published in 1930.[28] Most of the work was completed by April 1929, but enough data had been secured a year earlier for Plaskett's first paper on the galactic rotation problem.

Plaskett was already aware of some of Oort's thinking, because, owing to an accident at the telescope, he had taken an extended holiday overseas in 1927 and had met with the Dutch astronomer in Leiden. The next year saw Plaskett's paper in the *Monthly Notices*.[29] At that point, Plaskett and Pearce had radial velocity measurements for nearly 400 early B-type stars plus some 600 previously measured velocities for O- and early-B stars from other observatories. Plaskett had Pearce and S.N. Hill make graphs and through a winnowing process cut the number of useful stars down to 549. Solving Oort's equation with his data, Plaskett found the location of the centre of the system at galactic longitude 324.°25 ± 1.°8, close to Shapley's and Oort's values. The constant in Oort's equation, given by him as approximately 0.019 km/s/pc, was recomputed by Plaskett from O and B stars of known parallaxes as 0.0155 ± 0.007 km/s/pc, a close agreement. Plaskett and Pearce published their observational data in the *Publications* in 1930 and in 1934, employing their final data for a new solution.[30] Their new value of Oort's constant was almost identical to the preliminary figure. The new calculations gave the solar velocity with respect to the galactic centre of about 275 km/s at a distance of 10 kiloparsecs. These values could be used to calculate the period of solar revolution and the mass of the galaxy.

Following up this work, a young Cambridge graduate, R.O. Redman,[31] who had gone to the observatory in 1928 as a temporary appointment, turned to a study of K-type stars between the magnitudes of 7.0 and 7.5 that lay north of declination −10° and within 10° of the galactic centre. His study[32] showed that the constant calculated for K-stars agreed with Oort's and Plaskett's values, but the location of the galactic centre was different, which he expected for late-type stars. Oort was pleasantly surprised to see such rapid confirmation. For Plaskett, it was a great personal triumph. Although he previously had garnered two honorary degrees and was FRS and FRSC, he now was elected vice-president of the AAAS, in 1928. In 1930, he won the coveted gold medal of the RAS, along with being named its George Darwin lecturer for that year. The Rumford Prize of the American Academy of Arts and Sciences immediately followed; and the Bruce Medal of the Astronomical Society of the Pacific in 1932 rounded out the major awards.

Despite the accolades, Plaskett and Pearce, like other astronomers studying galactic scale, realized that there was still a major discrepancy in the distances. To solve this, they turned back to the interstellar absorption problem that Plaskett had attacked nearly a decade earlier. Eddington's 1926 theory had some points in common with Plaskett's earlier notions but assumed that interstellar calcium was spread in a cloud throughout the galaxy and that the stars moved through it as though the gas were relatively stationary. The stars showing calcium absorption would have to be earlier than B3 because of their high velocities, great distances, and sharp intrinsic calcium absorption lines. Struve, who had opposed Eddington's ideas earlier, was won over after further investigation. A paper by him and B. Gerasimovic in 1929 argued that the interstellar calcium also rotated with the galaxy and that their measurements suggested that the rotational constant for calcium was just one-half that of the early type stars. The data on those stars amassed at Victoria were well suited to test Eddington's hypothesis. Plaskett and Pearce had the material in hand before Struve and Gerasimovic's paper was published, and it has been suggested that Struve might have learned about it from Plaskett himself.[33]

Plaskett and Pearce's study of the problem appeared in 1930.[34] An analysis was made of 261 O- and B-type stars with Ca II lines. By two methods of calculation, they obtained rotational terms for the gas of 7.3 km/s and 7.9 km/s. The centre of the gas motion was estimated to be at galactic longitudes 335.1° and 331.7°, respectively, both differing from the centre of the stellar system. Eddington's theory was confirmed, as Plaskett and Pearce found that the velocity ratio of stars to gas was 2:1 for any group of stars chosen. A more complete study was made with the complete data and published the following year.[35] The 2:1 ratio was again confirmed, and Pearce was able to find a case of Ca II in the spectrum of a star later than B3. This was Plaskett's last major contribution.

The work of Plaskett and Pearce on galactic rotation and interstellar absorption, while drawing the most international acclaim, was only part of the quickening tempo of the observatory. Until the mid-1920s, the DAO had only four full-time astronomers and the occasional summer assistant or visitor. By the mid-1920s, however, both permanent and temporary staff grew. In 1926, Aitken approached Plaskett about taking on H.F. Balmer, one of Chant's graduates, who was working at Lick.[36] Plaskett was uncertain he wanted him but did negotiate with the ministry in Ottawa concerning a new full-time position. This was authorized. But Balmer took a position at Michigan, and Plaskett made an excellent alternate choice in Carlyle S. Beals (1899–1979), a recent London graduate.[37] A Nova Scotian, Beals was educated at Acadia University, spent some time at Yale, and took an honours degree in physics from Toronto, where he studied with J.C. McLennan. From Toronto, he went to Imperial College to work with noted laboratory astrophysicist Alfred Fowler. He obtained his PH D in 1925, and it was through Fowler's recommendation that Plaskett took him. Plaskett set him to work on the Wolf-Rayet stars, to which he himself had alluded in his earlier review of the problems of o-type stars.

Beals's first paper on Wolf-Rayet (WR) stars appeared in 1929.[38] He had obtained plates of all known WR stars visible from Victoria, to which he added Plaskett's plates and southern hemisphere plates provided by Shapley. These hot stars are anomalous in that they have, in addition to absorption lines, broad emission lines. In his early work, Beals revised the Harvard classification of WR stars based on the intensity ratio of He I to He II lines and hypothesized that the emission is caused by the ejection of atoms by selective radiation pressure. He also called attention to the possible relationship of WR stars to P Cygni stars, which have narrow emission lines, and novae. Beals divided WR stars into two classes, WC, with strong carbon lines, and WN, with strong nitrogen lines.[39] He was soon aknowledged an expert on WR stars, P Cygni stars, and similar objects by the astronomical community, having served on the IAU's committee for WR stars from 1933[40] and being invited to address an international symposium in Paris in 1939.

Besides the intensive research effort of the Plasketts, Harper, Pearce, and Beals during this period, the observatory provided a practical training ground for astronomers. Student assistants, especially in summers, carried a reasonable burden of the work, many of them contributing to observatory publications. W.H. Christie remained at the observatory until 1928, when he moved to Mt Wilson, and returned briefly to the observatory 1937-9. Peter M. Millman, a graduate of the Canadian Academy in Kobe, Japan, was an assistant in 1927, while a student at Toronto. Two University of British Columbia (UBC) students, later to become prominent members of staff, began work as summer assistants: Robert M. Petrie (1906-1966)[41] and Andrew McKellar (1910-1960).[42] In addition to summer assistants, faculty members of Canadian and nearby American universities were welcome visiting researchers, including A.V. Douglas and J.S. Foster of McGill and J.W. Campbell and E.S. Keeping of Alberta.

Plaskett was able to obtain excellent replacements for staff members who departed. His son, Harry, moved to Harvard in 1928 as a lecturer; in 1932 he was called to Oxford to become Savilian professor of astronomy. Taking his place was R.O. Redman. Redman was temporary in 1928, permanent the following year. His work in extending Plaskett and Pearce's galactic rotation studies impressed Plaskett, who described him as a 'splendid astronomer,' but in 1931 Redman was appointed to the Solar Observatory at Cambridge as F.J.M. Stratton's assistant.

During this period Harper continued his steady output of spectroscopic binary orbits, and many of the student assistants cut their teeth in this field. Staff member S.N. Hill was also involved in this line until his retirement because of ill health in 1934. The possibility of new appointments during the Depression was slight, but Plaskett managed to hire Frank S. Hogg (1904–1951),[43] a former Chant pupil and Harvard University's first astronomy PH D, as Redman's replacement in 1931. Hogg immediately joined in the radial velocity work. The observatory received a bonus in Helen Sawyer Hogg, a recent Radcliffe PH D, the first to be supervised by Shapley at Harvard in his field of variable stars in globular clusters. She had married Hogg in 1930 while in Cambridge and went with him to the observatory, although she had no official appointment. Through the kindness of Plaskett, telescope time was made available to her to continue this important new line of study. She was the first to seriously employ direct photography at the Newtonian focus of the 72-inch telescope, which had always been used for spectroscopy.[44] The Hoggs departed for the University of Toronto in 1934, however, in anticipation of the opening of the DDO.

The Depression had other, negative effects: salaries of the astronomical staff were abysmally low and were not ameliorated for some time, although Plaskett and others believed that the work itself was compensation. This was not a peculiarly Canadian problem, though, for as Plaskett's friend R.G. Aitken remarked in 1933, 'We are having a lot of fun just now with bank holidays and so on. Happily, astronomers never have any money anyhow, so we are not very seriously inconvenienced.'[45] Other minor irritations came in departmental penny-pinching in Ottawa which led to fights over whether the observatory could have an automobile or whether frayed cables supporting the observing platform could be replaced.

Plaskett had been due to retire on 31 July 1935, but the government allowed him six months' leave with pay. He left the directorship in January 1935 and, after a visit to England, where he was Halley lecturer, retired to Esquimalt, BC. He continued some work of his own and was appointed a consultant for the construction of the new 82-inch telescope at the McDonald Observatory in Texas, his last serious effort. He died in Esquimalt in 1941. Seldom has one man had such a great effect upon a science in a country as he, for J.S. Plaskett, more than anyone, had assured Canada's place in the international astrophysics community.

The Second World War disrupted activities at the DAO, as it did that of most observatories, and the majority of staff became involved in war work. Calgarian Walter H. Stilwell, a doctoral student at Berkeley, became an astronomical assistant in 1939, replacing Guy Blanchet, who had worked at Victoria from 1936. Joining the staff during the war were Kenneth O. Wright, a Toronto graduate, and in 1943 Jean K. McDonald, from the University of Alberta. Pearce became a local army instructor, while Wright and McKellar crossed to Vancouver to teach officers at UBC. Beals and Petrie were heavily involved in civil defence in the province. Petrie and McKellar were seconded to the National Research Council (NRC) in 1944, then to operations research for the Royal Canadian Navy. Stilwell worked for the Geodetic Survey in airfield surveys and left the observatory in 1945 to pursue surveying.

Despite disruption, work went on during the war. Pearce studied the Pleiades, continued the B-star program and, with Stilwell and Petrie, kept the spectroscopic binary work alive. Beals continued his studies of WR and P Cygni stars until his departure for Ottawa in 1946. Wright was involved in calculating curves of growth. An important visitor during the war was Anne B. Underhill, a UBC MA, who studied the Stark effect in B-type stars. E.S. Keeping from Edmonton also accomplished some spectroscopy during the war. When the situation in Victoria returned to normal in 1946, with staff returned, a long period of slow growth and diversification of research interest resumed.

As Plaskett had prophesied, Harper succeeded him in the directorship, but his tenure was short-lived. He travelled to Stockholm for the 1938 meeting of the IAU and fell ill, requiring hospitalization in Rostock. With the war coming on, Pearce, the second-in-command at Victoria, asked Vincent Massey, then high commissioner in London, to bring Harper and his family out of Germany.[46] This was accomplished in time, but Harper never recovered, and he died in 1940. His output was astonishing. Although his work was almost entirely restricted to binary stars and spectroscopic parallaxes, he published some 100 papers while at Victoria and had obtained 7,000 plates. He had computed more then 100 orbits which, at the time of his death, accounted for about a quarter of all orbits in the world.[47]

Pearce succeeded Harper, but as 'Head Astronomer,' since, with the dismantling of the Department of the Interior and the substitution of the Department of Mines and Resources, the term *director* was dropped. Beals was his assistant. When J.S. Plaskett retired, his position had been filled by two men, Petrie and McKellar. Petrie had finished his undergraduate work at UBC and continued his studies under H.D. Curtis at Michigan, where he received his PH D in 1932. McKellar, two years behind Petrie at UBC, proceeded to the Lick Observatory and took his doctorate at Berkeley in 1933. After a year at the Massachusetts Institute of Technology, he went to Victoria.

At first, both Petrie and McKellar were drawn into the usual programs of the observatory, publishing orbits of spectroscopic binary stars. Since Harper's interests were comparatively narrow and Pearce's wider-ranging than Plaskett's, Petrie became Plaskett's true heir at the observatory. They were similar in their administrative capabilities, both competent instrument designers, and both committed to the study of the early-type stars. Petrie made a close study of o-type stars just after the Second World War;[48] his comprehensive program for the B stars, continuing Pearce's work, was a product of the 1950s. His 1949 study of the o-type stars established the temperature scale within the Harvard classification of stellar spectra.

McKellar, like Beals, was more of a laboratory astrophysicist than an observer. He soon established a reputation in the study of spectral lines intensities, molecular bands, and research into the nature of molecular bands in the stars, planets, and comets. He calculated lithium isotope ratios from lithium bands in stars and the intensities of carbon bands in R stars. But he, like other members of the staff, was drawn to the problem of the interstellar medium. Plaskett and Pearce were contributors to the study of interstellar calcium, and Beals was the first to discover that calcium lines in stars could be double, thus confirming that interstellar calcium was not uniformly distributed in discrete clouds moving at differing velocities.[49] Interstellar lines arising from atoms, such as titanium, iron, sodium, calcium, and potassium, were noted by the 1940s. There were, however, several lines that defied interpretation.

In 1940, McKellar argued that some of these sharp lines might be 'skeletons' of molecular bands, the lines arising because, at the temperatures probably available in space, the ground state transitions predominated in atoms and should in molecules as well. He tentatively identified four of these lines as diatomic hydrides.[50] This was confirmed almost immediately by Adams at Mt Wilson, who discovered both CH and CN interstellar lines. McKellar's view was also substantiated in the laboratory in 1941 by Gerhard Herzberg and A.E. Douglas at the University of Saskatchewan, when they identified three of Adams's lines as the CH+ radical.

A year later McKellar pursued the question, publishing a comprehensive study of diatomic molecular spectra.[51] In this paper he calculated the energies available to diatomic molecules which give rise to interstellar lines and found a 'rotational' temperature of 2.3 K. This value was remarkably close to the 2.9 K temperature of the universe derived from studies of the microwave background radiation discovered in the 1960s.

Another important aspect of the work at Victoria was the interest in instrumentation. Both Plaskett and Young were knowledgeable in mechanical and optical design; the younger generation – Beals, McKellar, and Petrie – carried on this tradition. So well designed was the 72-inch telescope that astronomers in California consulted the Victoria staff freely. When the Lick Observatory was modernizing its Crossley reflector, the director, W.H. Wright, inquired about the use of a worm-gear for tracking; Harper was able to reply that after twenty years' use there was no tracking error due to the worm.[52] At the same time, F.G. Pease was at the observatory to confer on the drive system of the 200-inch telescope, then being designed. Beals, in the meantime, had designed a new self-recording microphotometer. Since photoelectric photometry was still relatively new, he had to obtain all the parts from the United States.[53] The original three-prism spectrograph had been designed for studying radial velocities, but Beals required a faster instrument with higher dispersion. In 1936, a Littrow-type grating spectrograph with a new aluminum-on-pyrex grating by R.W. Wood was built to the specifications of Beals, Petrie, and McKellar. Another, similar grating was obtained the following year.[54] The mechanical work was undertaken by S.S. Girling of Victoria, who built most of the observatory's new instruments during the 1930s and 1940s. More advances followed: Beals had a semi-automatic intensitometer built; Petrie designed a projection micrometer in 1937; and experiments in aluminizing the secondary mirror began in 1938. By 1943, silvering of mirrors was abandoned at Victoria, as the reflectivity of aluminized mirrors was substantially greater.

When one looks over the bulky bound volumes of the DAO's *Publications* from the early 1920s to the early 1950s, and reflects upon the extent of first-rate research they hold, it seems difficult to believe that it was achieved by a handful of astronomers and one telescope. Unlike the great astrophysical centres further south along the Pacific coast, Lick and Mt Wilson, the DAO did not possess a battery of large instruments, the backing of a Carnegie Institute, or the relationship with renowned scientific schools like the University of California or the California Institute of Technology. Successive government departments in far-off Ottawa provided just sufficient funds for efficient operation and just enough staff members to keep the telescope in full operation. In some ways, as C.S. Beals and J.A. Pearce have related to me in interviews, the resulting obstacles were actually turned to their advantage. Because the Ottawa bureaucracy was so distant, the directors, Plaskett, Harper, and Pearce, could operate the observatory with little interference from R.M. Stewart or deputy ministers.

The DAO's tremendous research output owed much also to the people who made up the staff during the first twenty-five years. Some observatories could be cauldrons of political intrigue and personal jealousies;

Donald Osterbrock has recently recounted the sordid early days of the Lick Observatory,[56] but such rivalry was never a part of the day-to-day life at Victoria. J.S. Plaskett hand-picked his staff precisely for their research capabilities: if they were not cut out for rigorous work, they were not hired. Both he and William Harper were, by all accounts, hard-working but straight-laced individuals, whose lack of a sense of fun was compensated for by a drive for perfection. Indeed, both are near caricatures of Victorian, small-town Ontario Protestants. None the less, no young man, fresh from university, could help but be influenced by their attitudes and their examples. Staff members had good connections with colleagues both east and south but typically relied on one another for intellectual sustenance.

The interests developed by the Victoria staff went well beyond the original, almost routine investigations inaugurated by Plaskett and Harper in 1918. Plaskett's view, that a national observatory ought to undertake routine but essential work, echoed William King's ideas about the role of the DO. But when that routine work, the collection and measurement of thousands of plates for binary star orbit calculation or for absolute magnitudes, could be turned to the exciting questions of galactic research – the nature of the interstellar medium, the rotational dynamics of the galaxy, and the nature of massive and peculiar stars – then the observatory could not help attracting bright young graduates. Plaskett's administrative genius, coupled with his natural desire to curry the attention and approbation of foreign astronomers, meant that the new staff members would be encouraged to take up new lines of research. Thus the research output of the observatory until well after the Second World War was a balance between routine and the frontier. This view of the role of a national astrophysical observatory survived long after Plaskett's retirement, being as much the credo of R.M. Petrie.

The DAO was built as a symbol of Canada's maturation as a nation, but its importance in the astronomical community was ensured by the appointment of Plaskett as first director. To a very large extent, the international importance of Canadian astronomy before the Second World War was centred on the Victoria observatory, having moved westward from Ottawa with Plaskett, Harper, and Young. The observatory's national influence was even greater: it was, until almost the 1960s, the research institution that formed Canadian astronomers. Virtually all of the graduates interested in astrophysics after the First World War worked at the DAO at one time or another, as staff members, visitors, or students. More important, the first graduate centre for astronomy, the University of Toronto, owed much of its early success to the DAO. When the David Dunlap

Observatory opened in 1935, it had R.K. Young as director and the Hoggs as staff members, all Victoria alumni.

Not surprisingly, the new Richmond Hill telescope, similar in many respects to the Victoria 72-inch reflector, was employed in programs virtually identical to those under way in the older institution. While the new observatory itself owed its existence to the long campaign of C.A. Chant, its research orientation lay in the long shadow of J.S. Plaskett.

CHAPTER SIX

Universities and Associations

The basis for modern Canadian astronomy was created between 1905 and 1945. The era was dominated by three major institutions: the Dominion Observatory (DO; see chapter 4), the Dominion Astrophysical Observatory (DAO; chapter 5), and the David Dunlap Observatory (DDO), discussed in this chapter as part of university astronomy. Equally important was the influence of their founders, William King, J.S. Plaskett, and C.A. Chant, respectively. All the features of contemporary Canadian astronomy – government institutions and research programs, university facilities, undergraduate and postgraduate education, professional and amateur societies, publications, connections with international bodies, and public dissemination of astronomy – were formed in the first half of the century. The period is characterized by quiet, incremental steps, rather than the more explosive growth of recent decades. That all of this could be accomplished was no mean achievement, considering the relative size of Canada's scientific resources, and that twenty of the forty years were spent at war or in the midst of depression.

One of the curiosities of the development of Canadian astronomy was the emergence of national institutions for astronomers but no provision for their training in Canada. Nineteenth-century astronomical education was limited to superficial arts courses, rarely accompanied by practical work. Astronomy as a subject for undergraduates had actually disappeared from the programs of most students by the turn of the century and would wane further in the new century. The PH D, for any science, much less astronomy, was late in taking root in the country; in fact, no PH D program in astronomy appeared until after the Second World War. A few universities, notably Toronto, Queen's, and Western, did provide sound undergraduate training. People hired by government observatories and university establishments, almost all of whom were Canadians, had to be trained abroad or be content with lower degrees in allied subjects from universities at home. In practice, those who took doctorates in astronomy or related subjects, such as mathematics or physics, typically did so at only a few, select institutions. Both Plaskett and Chant had friends at the Lick Observatory and tended to counsel students to attend the University of California; Chant, who had taken a doctorate at Harvard, also recommended his alma mater. A few others, perhaps because of the still-strong ties to Britain, left for London. On the undergraduate level, however, the centrality of Toronto cannot be overemphasized in this period, as a list of astronomy graduates of the first half of the twentieth century shows.[1]

In this same period, the RASC grew steadily in size and in geographical representation, with the addition of centres from one coast to the other. This amalgam of professionals with amateurs was successful because it suited the needs of a small population. The society's *Journal* and *Handbook*, both edited by C.A. Chant, gained early international recognition. The *Journal*, being a vehicle for professional articles, provided a window for the international community into Canadian endeavours. International co-operation by Canadian professionals grew alongside the society, through attendance at meetings first of the American Astronomical Society, later of the IAU and organizations such as the Astronomical Society of the Pacific. Participation in American societies did not, however, draw Canadians into the US astronomical community. The size and close-knit character of Canadian astronomy before 1939 ensured close co-operation within the country.

Until quite recently, the story of the training of professional astronomers in Canada is simply the history of the Department of Astronomy, University of Toronto.[2] American and British universities played supporting roles, but Toronto's domination of astronomical education was almost complete. There, as in the RASC, we see the considerable influence of Clarence Augustus Chant. As a research astronomer, Chant would not be remembered today, but his organizing abilities, his gifts for teaching and popularization, make him, together with J.S. Plaskett, the co-founder of modern Canadian astronomy.

Chant graduated with a BA in physics in 1890 from Toronto. After a year in the civil service, he joined the physics department at Toronto in 1891, being made a permanent lecturer the following year. He had had the same teachers as Plaskett and J.C. McLennan: Alfred Baker in mathematics, who taught elementary astronomy, and James Loudon in physics. Loudon, who had built up the physical laboratory – with Plaskett as his mechanical assistant – expected his students to master experimentation. Indeed, few of his students developed into theoreticians. Chant's early work was chiefly optical. He published an article on the reflectivity of glass and mirrors in the *Transactions* of the Toronto Astronomical Society in 1904 and read a number of papers to that society on physical subjects. Armed with a Harvard PH D in physics, which he had taken a leave from teaching to obtain, Chant began to think about the introduction of astrophysics – then a relatively new academic subject – into Toronto's curriculum. He first persuaded the mathematics and physics departments, as well as the university senate, to approve in 1904 a new fourth-year honours option in astronomy. The year was to be devoted to advanced traditional astronomy and mathematical methods.

As his own training in astronomy under Baker had been more or less of the armchair variety, he wrote to W.W. Campbell in 1906 saying that he would like to spend the summer of 1907 at the Lick Observatory to learn the latest techniques.[3] Plaskett, when he visited Lick that year, also mentioned the possibility of Chant's visit. Campbell aceeded to his request, and Chant remained at Mt Hamilton from June to September 1907. Although Lick was known best for radial velocity and binary star work, it was equipped for a variety of research, including meridian observations and laboratory astrophysics.

On his return to Toronto, Chant established a course in physical optics and an introduction to astrophysics in the 1907–8 school year. The course was intended for upper-level honours undergraduates. Because of the path of development of the sciences at Toronto, elementary astronomy and celestial mechanics were in the hands of the mathematics department. Although Chant was promoted from lecturer to associate professor the following year, and his first students – Harper, Motherwell, and Young – were working towards their degrees, he felt that astronomy at Toronto required more publicity. To this end, he tried to secure funds for a visit by Campbell for lectures on the latter's radial velocity work.[4] As he intimated to Campbell, 'Your visit would help me in my fight for the recognition of Astronomy here ... As the years go by I hope to hand over to you for completion of their astronomical training some of my best students.'[5] Unfortunately, funds could not be secured and Campbell did not visit. But Chant kept his promise: Young was sent to the University of California as a fellow in 1909.

In 1909, Chant added two laboratory classes, one in astrophysics and one in practical computation and observational astronomy, as an adjunct to the mathematics department's course on practical astronomy. As the university possessed no observatory of its own, students were allowed limited access to the Toronto Meteorological Observatory, then on campus, where the 6-inch Cooke refractor and 3-inch transit afforded some observational practice. By 1910, the physics department had obtained a Toepfer measuring microscope for measuring spectrograms. Plates were provided by Plaskett and by Sebastian Albrecht of Lick.[6] Lectures on campus in 1911 by R.G. Aitken, Lick's second-in-command, must have had a salubrious effect. When the 1913–14 year began, astrophysics had been made into a sub-department of physics. Chant was the only faculty member, but recent students Ernest Hodgson and I. Pounder were named assistants, along with three class assistants. Over the next few years, several of the acting assistants went on to graduate study and into astronomy as professionals, including J.P. Henderson, H.H. Plaskett, and J.A. Pearce. Chant had negotiated with the mathematics department and with the university president in 1912 to have another staff member appointed and recommended R.K. Young, who had just received his PH D from Berkeley.[7] Nothing came of the proposal, and Young went to Kansas.

The second student sent to Lick, R.S. Sheppard, did not turn out as well as Young. Campbell would not keep him on, as he had, in Campbell's opinion, no independence of thought and very poor English. Chant's reply to Campbell's chiding admitted: 'From the first year there is close specialization, and so when our men go abroad they usually make good "researchers," but for my part I would like to see more "culture" insisted upon.'[8] None the less, the astronomical curriculum remained intensively research-oriented for years to come. Not all students had turned out like Sheppard; R.J. McDiarmid (BA, 1910) had gone to work at the Allegheny Observatory as a fellow, then to Princeton, where he completed a PH D in 1915. Harper, Cannon, Motherwell, and Parker all took the MA at Toronto after completing their prescribed courses of study.

In 1916, Chant applied for promotion to full professor and wrote to both Campbell and G.E. Hale for references. Although President Robert Falconer was sympathetic, the promotion did not become official until 1925. The end of the war and the return of students did coincide with an important change: the creation of an independent Department of Astronomy in 1918. By this move, all astronomical courses in the Faculty of Arts were grouped under Chant's supervision. The curriculum, with only minor changes and reshufflings, stood for the next thirty years. From mathematics came courses on descriptive astronomy, elementary astronomy, advanced astronomy, and elementary and advanced practical astronomy; from physics came introductory astrophysics and the laboratory courses; from the sub-department of mechanics came celestial mechanics and least squares. These were reorganized in 1920 into pass, pass and honours, and honours courses.

While Chant was not primarily a practising astronomer, he did undertake two eclipse expeditions before the arrival of Young. In 1918 he journeyed to Matheson, Colo, on behalf of the RASC, taking with him the 4.5-inch Cooke refractor loaned to him by the DO; Klotz was clouded out in nearby Denver.[9] More important was the eclipse expedition of 1922 to Wallal, Western Australia. Chant, with his wife and his daughter Jean, joined the Lick expedition led by Campbell, taking with him a specially built telescope – the 'Einstein Camera' – built with the aid of Prof C.R. Young of the Engineering Faculty. R.K. Young, by then at Victoria, was allowed by Plaskett to participate, and it was he who measured the plates to confirm that Einstein's predicted shift in stellar positions occurred.[10] During this period, Chant's agitation, recounted below, for a large observatory quickened, although the results were still a decade away.

The 1920s saw slow growth in astronomy at Toronto and a continuing trickle of students destined for professional astronomy. J.A. Pearce, recently returned from Europe, completed his undergraduate work, taking his BA in 1922. He then transferred to the Lick Observatory. H.F. Balmer (BA, 1924) also studied at Lick, went on to the DO, and by 1931 was working at the Buffalo Museum of Science. In 1925, when Chant was finally promoted, his long-awaited assistance was provided with the appointment of R.K. Young as associate professor. The second generation of Toronto astronomers was trained over the next few years. Frank Scott Hogg, an RASC gold medallist, graduated in 1926, moving to Harvard University, where Harlow Shapley was building an important centre for astrophysics. Hogg took his PH D in 1929 and joined the staff of the DAO in 1931. He was followed by Peter M. Millman, another RASC gold medallist (1929), who also chose Harvard for graduate studies. Millman had already shown an interest in meteoritic astronomy and continued in that relatively uncultivated area both at Harvard and at Toronto, where he was appointed a lecturer in 1933.

The growth of astronomical education at Toronto has a story within a story, the development of the DDO. Clarence Augustus Chant was at the centre of that drama, as well. Chant has provided a detailed account of how the observatory came into existence, and R.K. Young has provided technical details;[11] we need only review the main steps. Chant was aware, when he began lecturing at Toronto, that teaching observatories are essential adjuncts to academic programs. He knew also that such observatories were typical features of American universities, but not of Canadian.

When the astrophysics program was under way, not one Canadian university actively operated an observatory, although several possessed one. Chant's agreement with the Meteorological Service for use of the Toronto Observatory instruments was a stop-gap at best. In 1905 he toured a number of eastern American institutions for ideas, but not until 1911, when a suitable plot of land in Toronto (at Bathurst and St. Clair) struck him as an ideal site for a 'high-class observatory,' did his plans begin to mature. Negotiations were begun with the city of Toronto in the same year, interrupted by the war, and reinitiated in 1919. In 1915 Chant had had architectural drawings made of the proposed 'Royal Astronomical Observatory, Toronto,' but had no obvious source of funds for building and equipping such an institution. The city agreed, in 1919, to allow the use of a park for an observatory, with the municipality maintaining the grounds and the university the buildings; the public and the RASC would have access to the observatory. Plans for the observatory were fairly definite, Chant proposing to obtain a 24-inch refractor and a 9-inch instrument for amateurs. Ralph De Lury suggested a solar telescope, and G.E. Hale, the master of observatory foundations, concurred.[12] Chant then hit the true snag: no money.

The sciences have received little private philanthropy in Canada, and what largesse they got went elsewhere than Toronto. The city's wealthy families were unable or unwilling to provide Chant with funds, but in 1922 he happened to be approached by financier David Dunlap, whose interest in astronomy had been aroused by one of Chant's lectures. This interest extended to the observatory project, but Dunlap did not actually commit himself to it financially. By that time Chant had decided that the traditional, medium-sized refractor would be less successful as a research tool than a large reflector, although the cost would be higher. In the interim, Young, who joined the department in 1924, suggested building a more modest reflector at the university. After inquiries, Chant and Young found that a 19-inch pyrex disc could be provided cheaply by Corning, and a blank was ordered. Young spent his spare time over two years and $1,500 to finish the equatorially mounted and clock-driven telescope. The gearing system was the same as that in the 72-inch at Victoria, which he had used for some years. When he reported[13] on the project in 1930, he still had no place to employ it.

Although David Dunlap had died in 1924, Chant decided to bring his new proposal for an observatory with a large reflector to the attention of Jessie Donalda Dunlap, Dunlap's widow. This he did late in 1926; Mrs Dunlap was enthusiastic about providing an astronomical memorial to her husband, and over the next two years details were worked out, including the choice of a site south of Richmond Hill. The site was, of course, far superior to the original park location, long since engulfed by the city; ironically, the Richmond Hill site, then some distance from the city, has in its turn suffered the same fate.

Chant had already discussed the idea of a large reflector with Sir Charles Parsons of Grubb-Parsons in Britain and, after Mrs Dunlap gave final approval for financing the project in May 1930, ordered a 74-inch reflector. This was the first large telescope constructed by Grubb-Parsons and a prototype for later instruments in South Africa, Australia, and Egypt; the design was by Chant, Young, and Cyril Young of Grubb-Parsons. The dome, building, and mounting were built in England and then re-erected in Richmond Hill by Dominion Bridge Ltd. The Administration Building, completed in 1933, finally became the site of Young's 19-inch telescope and, eventually, the old Cooke 6-inch telescope of the Toronto Meteorological Observatory. The glass disc for the 74-inch was more difficult, and only when Young heard about Corning's plan to cast a 200-inch disc for Hale's new observatory did a workable alternative present itself. Corning agreed to cast a 74-inch disc of pyrex as a trial run for the 200-inch, which was accomplished in the fall of 1933. The disc was shipped to Grubb-Parsons's Newcastle works and was ground and polished under Cyril Young's supervision; it arrived at Richmond Hill in May 1935.

The completed instrument operated in two modes, as an f/4.9 Newtonian and as an f/18 Cassegrain. Almost all the early work with the instrument was meant to be done at Cassegrain focus, as at Victoria, with a spectrograph built by Hilger in England. The new instrument, second largest in the world, made a substantial difference to the university's research.

Until the beginning of the 1930s, the academic staff had consisted only of Chant and Young; in anticipation of a major research instrument, the university was persuaded to increase the size of staff. In 1933, Peter Millman joined the staff, and in 1935, when the observatory officially opened, Frank Hogg was brought from Victoria as a replacement for Young's position, vacated when he became head of department and director of the DDO. Chant, now seventy, retired on the opening of the observatory. Helen Sawyer Hogg was named a research associate, while John F. (Jack) Heard, a recent London PH D with Fowler, returned to Canada to become the junior member of the faculty. This staff carried on the academic and research work until the war.

The research program at the observatory commenced immediately under Young's direction.[14] Carrying on his life's work, Young initiated a series of observations to obtain the radial velocities of 500 stars in Kapteyn's selected areas, north of declination + 15° and brighter than magnitude 7.59.

The plates were measured by departmental students Ruth Northcott, A.F. Bunker, and William Buscombe; the last later worked at Saskatchewan and in Australia. This series was complete by 1939, and an additional 600 stars were added for observation during the war. Frank Hogg and Jack Heard were the major participants with Young in the radial velocity program, while Millman carried on meteoric research and Helen Hogg continued her studies of variable stars in globular clusters. This last project, begun at Harvard and carried on at Victoria, was the only work done at the Newtonian focus of the 74-inch for many years. Following the lead of the DAO, Young established a *Publications* series for the observatory, along with *Communications* – reprints from other journals – to which Helen Hogg contributed the first item in 1938.

The undergraduate program had been reorganized for 1933–4, dividing the courses into pass, general, and honours streams, and the course titles remained in force, with a few changes, until the 1960s. A distinction was made between general students, who proceeded to the three-year ordinary degree, and the four year honours students, who were highly specialized. Most of the courses were traditional, but the newer staff members introduced some new courses reflecting recent developments, such as meteoritics by Millman and spectrophotometry by Frank Hogg. Toronto's lack of academic broadening, which Chant lamented in 1912, remained a defect until the 1969 revisions which, for astronomy at least, were largely cosmetic.

The growth of undergraduate enrolments before the war was healthy: in 1937–8, 42 students took the pass courses, 35 the general, and 38 honours, for a total of 115. The number of graduate students was not so encouraging. Only the MA was offered, but rarely was there a year with more than one student registered, and some years went by with none. The old practice of employing senior undergraduates as assistants remained, however, and a number of these and recent graduates were drawn into the work of both department and observatory. Some, like Ruth J. Northcott and Edna Fuller, became permanent employees.

The Second World War took a toll on both the academic staff and the work of the observatory. In 1940, Heard joined the Royal Canadian Air Force, as did Millman the following year. Also lost to the war effort was Gerald F. Longworth, the telescope's superintendent for some forty years. Promising students were also called away, and the general student numbers began to drop. In 1940–1, the department was teaching only 66 students; two years later, the number fell to 59.[16] Ruth Northcott joined the academic staff in 1944. Immediately after the war, graduate students returned and undergraduate enrolment jumped to 189 in 1946–7.[17] To compound the problem, Millman resigned to go to Ottawa and R.K. Young retired. Frank

Hogg became head and director in 1946. To fill the gap left by Millman, Ralph Williamson, an American theoretician and specialist in the new field of radio-astronomy, was hired in 1947. This began two new lines of research in the department that would eventually become important.

Higher education in astronomy east of Toronto scarcely existed until after the Second World War, but two important centres of activity should be noted, McGill and Queen's universities. Although McGill's original observatory was intact, now under the direction of the Department of Civil Engineering, it ceased to be of importance to astronomy. Partly funded by the Canadian Meteorological Service, the observatory's chief work continued to be in meteorology and timekeeping. Descriptive astronomy remained part of the curriculum at McGill for some years, yet no department of astronomy ever emerged. An abortive attempt to inaugurate an astronomy program was made in 1923 by A.H.S. Gillson of the mathematics department, who wrote to Chant inquiring about suitable courses, texts, and instruments; despite Gillson's teaching an extension course in astronomy in 1924, nothing came of the proposal.[18]

One of the scientific strengths of McGill was its physics department, staffed predominantly by Cambridge-trained men; among its early students was Alice Vibert Douglas (b. 1894). Douglas took an M SC under J.A. Gray, then studied with Rutherford, a former McGill professor, and Eddington at Cambridge for two years. She returned to McGill in 1923 to work on a PH D. In the summer of 1925 Douglas was a research assistant to E.B. Frost at the Yerkes Observatory; there she entertained the idea of ascertaining spectroscopic absolute magnitudes of A-stars and was able to obtain some 250 Yerkes plates for that purpose. After conferring with Shapley at Harvard, she was able to work up results for a PH D dissertation,[19] directed by Gillson, who was also McGill's expert on relativity theory. Already a demonstrator in physics, she was appointed, in 1927, lecturer in astrophysics, a post she retained until 1939. This appointment allowed the physics curriculum to expand to allow for an honours option course in astrophysics for third- and fourth-year and M SC students. Among those students were Miriam Burland, who joined the DO's staff in the 1920s, and J.F. Heard, who moved on to the University of Toronto.

At Douglas's insistence, McGill purchased a Moll microphotometer with which she was able to analyse spectra of Cepheid variables from plates obtained with Henroteau at Ottawa.[20] A noteworthy member of the physics department at the time was John Stuart Foster, who had produced the definitive laboratory studies of spectral line broadening due to the Stark effect, the broadening of certain lines of atoms in electric fields discovered by J. Stark in 1913. In 1931, Douglas convinced Foster that they might be able to apply his laboratory values for hydrogen and helium to the lines in the B-type stars. An NRC grant allowed them to spend part of the summer of 1932 at the DAO obtaining plates. Several joint papers resulted;[21] this study showed that in low-luminosity O-stars, the Stark effect accounted nicely for some helium line broadening, but not for some others, which, they claimed, required a departure from thermodynamic equilibrium. This research was extended and confirmed at Victoria several years later by Anne B. Underhill of UBC and William Petrie of the observatory.[22] As Otto Struve, another pioneer in the study of the Stark effect, noted later, these ideas were 'quite ingenious' but probably could be applied only to anomalous stars such as novae.[23]

In 1939, Douglas left McGill for Queen's ending the work in astrophysics at the former institution. Astronomy at Queen's, despite the presence of N.F. Dupuis and the relocation of the observatory to the campus in 1907, had languished. Indeed, the only notable graduate, S.A. Mitchell, later director of the Leander McCormick Observatory in Virginia, took his degree in 1894.[24] By the time Douglas arrived, only one course in descriptive astronomy for arts students was offered, by Prof Johnstone of mathematics. In 1943, Douglas took over the astronomy course and used the old Clark 6-inch refractor for student work. Only after the war did astronomy expand at Queen's, and then chiefly in radio-astronomy.

Besides Toronto and Queen's, the only Ontario university to take astronomy seriously before the Second World War was Western. Henry R. Kingston, who had taught at Manitoba and was active in the RASC centre there, went to Western in 1922 to teach mathematics. The London Centre, which he formed in 1922, was active and participated in the solar eclipse observations in Quebec in 1932. There was no observatory and no program, but in 1939 the widow of Canon Hume Cronyn presented a $40,000 bequest to the university for an observatory in her husband's honour. The small observatory, completed in 1940, was equipped with a 10-inch refractor by Perkin-Elmer,[25] but no research of consequences was undertaken until the mid-1950s.

Western Canadian universities arose later. Provincial universities were created in Alberta (1906) and Saskatchewan (1907), soon after those provinces were established from the old Northwest Territories. UBC, begun as an affiliate of McGill, and Manitoba, a federation of church colleges, both had fitful starts. Not until the 1920s could any of the four universities be said to be on solid footing.

Astronomy had a somewhat stronger place in the curriculum of the western schools compared with all but a few of the eastern universities, because science was considered an important, integral part of the programs of western schools. Yet in none of them did astronomy become a significant area of study. This may be explained, at least in part, by the strong practical orientation: towards agriculture, engineering, and other practical subjects necessary for provincial development.

At Manitoba, the mathematics department, which had taught descriptive astronomy from the turn of the century, became the Department of Mathematics and Astronomy yet, curiously, never became an important centre of astronomy in terms of research or graduates. By the time of the First World War, three members of the department taught the subject: the head, Neil B. MacLean, a distinguished veteran and Toronto graduate; Lloyd A.H. Warren, a Queen's graduate who obtained a doctorate at Chicago; and, from 1914 to 1921, H.R. Kingston, who had followed the same route as Warren. By the early 1920s, the department offered descriptive astronomy on two levels, along with practical astronomy and celestial mechanics. There was no observatory, and, although Warren was both a fellow of the RAS and a member of the American Astronomical Society, he seems to have made no important contribution to Canadian astronomy. One Manitoba physics graduate, Malcolm Thomson, notes that by the late 1920s celestial mechanics was rarely taught, since there was little demand for it.[26] The only other graduate of note, again not in astronomy proper, was A.G.W. Cameron, who completed a Saskatchewan PH D and went on to a distinguished career at Atomic Energy of Canada Limited, at Chalk River, and later in the United States. By the time he was a student at Manitoba, the astronomy offerings were reduced to descriptive astronomy and navigation.

At Saskatchewan, too, astronomy was under the wing of mathematics, where it was taught as a general course and as a more advanced mathematical course. A small observatory was located on the campus from the late 1920s, and the teaching was undertaken primarily by Alfred Pyke, a Toronto graduate, and Lloyd Dines, who had a Chicago doctorate. While traditional astronomical research went undeveloped at Saskatchewan, the area of laboratory astrophysics, then almost non-existent in Canada, appeared in the person of Gerhard Herzberg, who had left Germany and gone to the university in 1935 at the invitation of chemistry professor J.W.T. Spinks.[27] Already in Germany he had inaugurated studies of astrophysically important molecular spectra, particularly the electronic spectra of diatomic and simple polyatomic molecules. A small research group, which attracted students such as A.E. Douglas, was formed and lasted until 1945, when Herzberg moved to Yerkes. In 1946, he formed a new spectroscopy group at the NRC. In 1938, he had suggested that hydrogen could be detected in planetary atmospheres, a feat not actually accomplished until 1958.

In Alberta, astronomy did not begin to flourish in the universities until after 1945. At the University of Alberta, J.W. Campbell, who had taken a PH D at Chicago in astronomy and had served in the artillery overseas, was appointed in mathematics in 1920.[28] Although he maintained an interest in astronomy throughout his career – he was founding president of the Edmonton Centre of the RASC and eventually national president in 1947 – and wrote a textbook on mechanics, for decades he and his department offered only an introductory course. E.S. Keeping, who went to the department from England in 1929, taught the course for many years and did undertake some research at Victoria during the war years. The only pre-war student of note was Jean K. McDonald, from physics, who went to Victoria. The university possessed no telescope until 1943, when a local amateur, Cyril Waits, donated his 12.5-inch reflector.

UBC, which probably had the weakest astronomy offerings of the western universities, produced more notable graduates than any other western school, largely because of its strong physics program. Astronomy, part of the mathematics curriculum, was in the hands of Daniel Buchanan, who had gone from McMaster University to Chicago to take his doctorate in celestial mechanics under F.R. Moulton in 1911. Chicago provided all four western universities with doctoral graduates to teach astronomy. Buchanan taught descriptive astronomy from his arrival in Vancouver in 1921 until the 1940s; he retired as dean of arts in 1948. On occasion, he also offered a course in celestial mechanics. The physics department had obtained the services of Gordon Shrum, a student and co-worker of J.C. McLennan at Toronto, where he was co-discoverer of the auroral green line and the liquefaction of helium. Shrum was a supporter of astronomy in the university and in the RASC. Among students who went on to astronomical or related careers were R.M. Petrie, his brother William, Andrew McKellar, Anne Underhill, Arthur Covington, and George Volkoff. The last, who joined the Department of Physics in 1940, is best known in astronomical circles for his co-authorship with J. Robert Oppenheimer of the now-classic paper on neutron stars. Spectroscopy was an important subject in the department by the early 1940s, and William Petrie, McKellar, and K.O. Wright all taught there briefly. The proximity of the DAO – and the willingness of its directors to employ summer assistants from the university – certainly contributed to UBC's importance in astronomical education.

The RASC[29] played the pivotal role in creating the interaction between professionals and amateurs. It has a long and fascinating history, and here we need review only a selection of important events in its history, but the institution, unique in so many ways, should have a history devoted to it.

It is an example of the successful transition from local society to national organization.

While many men and women played important roles in shaping its early history, the stamp of C.A. Chant on the society's outlook and development is of paramount importance. Chant had been elected to membership in the Toronto Astronomical and Physical Society in 1892 but came into prominence in the 1900s. The society had become the RASC in 1903, though still a Toronto-based organization. In that year Chant became editor of its *Transactions*. The following year saw his election to the presidency, a post he held until 1907.

Before incorporation of the RASC a few, local astronomy clubs in Ontario were affiliated to the Toronto organization. These were all small and relied upon the enthusiasm and knowledge of one or two local people. Typical examples were the Meaford Astronomical Society and the Owen Sound Astronomical and Physical Society, the latter having twelve members in 1903. The impetus for such organizations was similar to that of the earlier mechanics' institutes, that is, as much social as intellectual. Inevitably, these small organizations, in small centres, could not live long when the key person died or moved. After the RASC become incorporated, with provision for affiliation of local centres with the national body, nearly all these local groups evaporated. The structure of a large, Toronto-based group was inappropriate for small, local interest groups, with slates of officers and councils. Indeed, such small organizations that do survive typically have a flexible organization, but such was not the case.

The first full-fledged centre of the society was that of Ottawa, organized in December 1906 at the instigation of DO staff members. Nearly all the observatory staff took an active part, and the meetings were held in the lecture room of the observatory. Talks by professionals, and King's policy of allowing the public to view through the 15-inch refractor on Saturday nights, attracted a good deal of attention. Since Ottawa had a large concentration of people likely to belong to such a group, including scientific civil servants and educated professionals, the centre was able to enrol 104 members by early 1907. The Ottawa centre flourished thereafter. The professional presence was, in the early years, greatest in that centre. In 1919, for example, 16 of its members were DO staff members, while at Toronto only Chant and J.A. Pearce were professionally trained, and at Victoria, only Plaskett, Young, and meteorologist Napier Denison were professionals.

An example of a local group that relied upon too few key figures is the Peterborough Centre, organized by Rev D.B. Marsh, a local amateur, in 1907. It initially enrolled a respectable 47 members. Not only did Marsh

lead the group, but his intellectual contributions sustained it. Several astronomers from outside Peterborough gave the occasional talk, but discussions and lantern-slides were more typical. By 1909, membership had
dropped to 24; by 1916, to 11. Marsh's move from Peterborough sounded
the death knell for the group, and it disappeared. A centre was formed in
Hamilton in 1908 and claimed 70 members the next year. In Regina, the
Saskatchewan Astronomical Society was formed in February 1910, with N.
MacMurchy as president; it was admitted as a centre in October. In the
following month, the Astronomical Society of Western Canada in Winnipeg, formed in 1909 by Prof N.B. McLean, of the University of Manitoba, was admitted as the Winnipeg Centre. In 1911, a Guelph Centre was
established, with Henry Westoby as president and about 75 members. With
the coming of war, however, those centres with no professionals to sustain
interest failed. In 1914, a Victoria Centre was founded that included members from the mainland – Vancouver did not form a separate centre until
1931 – but there, although there was no university, the promise of the
building of the DAO and the subsequent active participation of its staff
members guaranteed a healthy existence. The national society had 450
members by 1916: 141 in Toronto, 93 in Ottawa, 26 in Hamilton, 20 in
Winnipeg, 30 in Guelph, 11 in Peterborough, 16 in Regina, 83 in Victoria,
and 30 honorary. It was already evident that the centres in Guelph,
Peterborough, Hamilton, and Regina were fading; by the early 1920s they
all disappeared.

Unfortunately, because local astronomical clubs were rare in Canada,
and centres of the society existed only in a few cities and not always for
long, some amateurs had to face a lonely existence; still, a few transcended
this typically Canadian problem. One such amateur was Joseph Miller Barr
(1857–1911) of St Catharines, Ontario, whose work on variable stars and
binaries led to articles published in the *Journal* of the RASC, the *Astrophysical Journal*, and local newspapers.

It took a long time for activity in Montreal to surface. Despite its status
as Canada's largest city, Montreal's predominantly French-speaking population had little interest in astronomy, and the English-speaking community could not look to McGill for leadership, because, although it possessed
an observatory, it had no astronomy program. In 1918, preliminary discussions were held with a view to forming a centre, which came about in
May with the election of Mgr Choquette of Ste-Hyacinthe as president.
None the less, few francophones joined. The group did survive and later
relied upon McGill professors, particularly A.V. Douglas, after her arrival
as a lecturer in 1927. By 1930, the Montreal Centre was third largest in the
country with 114 members.

London, despite its university and population, saw little activity until H.R. Kingston began teaching astronomy at Western. In 1922, with Kingston as president, a centre was created. Hamilton's centre was re-established in 1930. No new centre was founded until the 1930s. Still, the growth of the society was very encouraging. From 450 members in 1916, it had nearly doubled to 738 by 1930.

When the RASC was first incorporated with a royal charter, it attempted to emulate its British namesake, the RAS, not only in its national and international aspirations but in its election of fellows and honorary fellows from overseas. Those who bore the title FRASC were those who had contributed to the society, such as John R. Collins, the long-time secretary, and distinguished amateurs, such as A.F. Miller, or professionals, such as King, Klotz, Plaskett, and De Lury. Honorary fellows included the foreign leaders of astronomy such as Hale, Campbell, Frost, and Pickering. In time, this practice died out, one suspects because being an FRASC never carried the weight of being an FRAS. Professional scientists, with or without titles, formed a vital segment of the society. For example, in 1909, of some 415 members in Canada, 87 or 21 percent, were scientists. Professionals took an active part in governance at both national and local levels: in 1930, national officers included astronomers Stewart, Kingston, Young, A.V. Douglas, De Lury, Harper, and Beals, along with some meteorologists, geophysicists, and physicists. Nearly every centre had a scientist either as the president or honorary president. In some centres, such as Toronto, it became a custom to alternate between amateurs and professionals in high offices.

The national society had, as its primary purpose, the encouragement of astronomy in Canada, but from early on it had international aspirations. In 1908, Chant and G.E. Hale corresponded concerning the admission of the society as a constituent member of the International Union for Co-operation in Solar Research, Hale's forerunner of the IAU. Hale wrote to support Chant's bid to involve the society,[30] and the latter was able to report that the council agreed, in February 1909, to participate and would appoint either Plaskett, Chant, Louis B. Stewart, or Sir Frederick Stupart as commissioner.[31] As we have seen, Plaskett became the representative at Pasadena for both the society and the DO.

Chant realized early that the society would prosper if it had means of communicating with its far-flung membership. In addition, it needed to establish credibility in the international astronomical community. To this end, discussions were held in 1905 towards creating a bi-monthly journal that could serve both the scientific needs of the professionals and the requirements of the society and its amateur members. The *Transactions* were allowed to lapse in 1905, and the *Journal* of the RASC was planned. In 1907, the first issue appeared. Reflecting the nature of the society, the *Journal* was a unique document. The editorial policy was stated by Chant in the first number: 'There are few technical articles on astronomy, which, if clearly written, have not a real value to the amateur, while the work of the latter is always of interest to his professional brother. There will be room for both in the pages of the JOURNAL.'[32]

In the Canada of 1907 the only outlet for scientific papers on astronomy, apart from official publications such as the *Report of the Chief Astronomer*, was the *Transactions of the Royal Society of Canada*, the pages of which rarely carried astronomical articles and were not readily available to non-fellows. There was no periodical for popular science. Canadian astronomers relied on foreign, purely technical journals as outlets, including the *Monthly Notices* of the Royal Astronomical Society, the *Astronomical Journal*, and G.E. Hale's *Astrophysical Journal*. The most readily available popular publication was the American magazine *Popular Astronomy*. The *Journal* of the RASC, combining as it did – and still does – technical articles, popular or semi-popular articles, and society news, differed from all of them.

The older *Transactions* was dominated by articles on physics. The new *Journal*, reflecting contemporary Canadian interest with accuracy, was, in the volume for 1908, dominated by geophysics and meteorology, and this orientation continued for some years. As Plaskett built up the astrophysics staff at Ottawa, more papers on radial velocities and spectroscopy began to appear. Mixed with this, however, were rambling literary and historical pieces, along with a number of astronomical poems, notable more for their enthusiasm than for literary merit. This, again, bore Chant's influence, for he delighted in anecdotes and curious news items and was a firm believer in a broad education.

In 1915, the *Journal* became monthly, although in practice it normally doubled up two issues so that 10 numbers appeared a year. It appeared in that format until 1946, when it returned to its current bi-monthly format. Chant, as editor, wrote a good deal of filler material. J.R. Collins was, for some years, responsible for assembling astronomical news and quotations from published papers. Officially, Chant had three associate editors, normally the directors of the two national observatories and the Meteorological Service. Thus, Stewart succeeded King as an associate editor, Plaskett was an associate for many years, and Stupart acted on behalf of the Meteorological Service. In effect, however, Chant seems to have undertaken most of the editorial work.

Chant began, as a service to amateurs, publication of a separate booklet listing astronomical events – the *Canadian Astronomical Handbook* – in 1907. After one further number, the almanac material was printed in the *Journal*, but the separate publication was revived in 1912 as the *Observer's Handbook*. Shortly thereafter, a British publication of similar content and identical name appeared, but Chant decided to retain the name. The *Handbook* has flourished ever since, despite a good deal of competition.

Until after 1945, the activities of the society tended to be centred locally, since travel between Canadian cities was still a major undertaking. The annual meeting was held in Toronto, since the national office and Toronto Centre's rooms were located together in the Royal Canadian Institute building. However, the annual meeting was more a Toronto event. Not until 1958, during the presidency of Helen Hogg, did the annual meeting move outside the city, to nearby Hamilton. Since then, the annual meeting, together with paper sessions in what has been termed the general assembly since 1962, has become peripatetic. This, of course, became possible only with the advent of rapid and relatively inexpensive air travel.

The society had participated as a corporate body in the Solar Union and had sent observers to Labrador for the 1905 eclipse; in 1932, a total solar eclipse was to be visible from Quebec, and the society dispatched an expedition to Louiseville, on the same site as the French expedition led by the Comte de la Baume Pluvinel. Chant did not accompany this group, but, together with Young, selected a site at nearby St-Alexis-des-Monts for the University of Toronto. The Hamilton Centre, with H.R. Kingston from London as its professional member, set up in Actonvale. Only Chant and Young were completely clouded out.

In these ways, and in the maintenance of a library and stock of instruments for the use of members, the society fulfilled its commitment to amateur astronomy. Apart from making the pages of its *Journal* available to professionals, the society aided the professionalization of Canadian astronomy by its creation, in 1905, of a gold medal, to be awarded to the student standing highest in honours astronomy at Toronto. Its recipients read like a 'who's who' of Canadian astronomy and have included W.E. Harper, the first recipient, in 1906, R.K. Young, F.S. Hogg, and P.M. Millman. In 1940, the Chant Medal was instituted for distinguished amateur contributions to the society, and in 1959 a Service Award was created for those who had aided the society but were ineligible for the other two medals.

During the 1930s, the society continued its slow growth and added two new centres, in Edmonton and Vancouver. In Edmonton, a group was formed in January 1932 by Prof J.W. Campbell, professor of mathematics at the University of Alberta.[33] He was able to obtain the requisite 50 members for centre status, and the organization was established in March. Local university scientists and visitors provided the public talks for the group. As with most centres of the society, initial enthusiasm cooled, and by 1937 the centre numbered only 35. The Vancouver Centre was formed at the same time under the direction of Gordon M. Shrum of the UBC physics department. A number of Victoria Centre members who lived on the mainland switched memberships, but both centres flourished.

Several centres built observatories or shared facilities with nearby universities. Edmonton had the use of the University of Alberta observatory, built in 1942 to house the reflector donated by centre member C.G. Waits. Montreal erected an observatory on Mount Royal near McGill. Most centres, however, did not obtain observing facilities until the 1960s and 1970s. The Toronto Centre had taken an early and active interest in Chant's proposal for a major observatory for the university, and it was at society meetings that Chant became acquainted with David Dunlap.

All the activities enumerated were those of English-speaking Canadians. In French Canada, the level of astronomical activity was much lower, but slow movement was taking place. In the nineteenth century, francophone interest in science had waned; for astronomy, only lectures in the collèges classiques and at the Université Laval continued. No professionals emerged. Even as the twentieth century opened, nothing further was done in astronomy at Laval or, after its establishment in 1920, at the Université de Montréal. During the 1920s and 1930s, however, educationists, particularly Adrien Pouliot, began to attack the older methods and curriculum. Slowly, modern science education grew in Quebec. None the less, there was amateur activity among French Canadians, though it was little known in the rest of Canada. In 1934, R.G. Aitken of the Lick Observatory wrote to Chant asking whether he knew of an observatory in Jonquière, Que; its owner had evidently ordered some lantern-slides from Lick and had failed to pay. Chant had never heard of it but found in Stroobant's list of observatories that one D'Antonio Poitras had an observatory that was said to have telescopes of 10- and 20-inch apertures as well as a 4-inch transit instrument.[34]

The Jesuits in Quebec were still interested in astronomy. The staff of the Collège Ste-Marie of Montreal observed the 1932 eclipse at Sorel with 6-inch and 4-inch refractors. This was reported to the *Journal* by Rev E. Cambron. It will be recalled that Laval had purchased a telescope and installed it on the roof of the Grand Séminaire in 1892. The telescope fell into disrepair and by 1941 had been nearly forgotten.[35] In that year, the Cercle astronomique de Québec, the Quebec City amateur group under the presidency of Paul-H. Nadeau, asked the university to move the telescope to the Martello Tower on the Plains of Abraham.[36] The rector, Mgr Camille Roy, was willing, and the Foucault reflector was duly installed in its new location. It remained there until the late 1960s.

Nadeau hoped to encourage interest in astronomy among young French Canadians. As he argued to Mgr Roy: 'C'est une occasion unique pour [l'Université] de prendre les devants dans le domaine de l'Astronomie que les Canadiens Français ont negligé par leur faute.'[37] The Cercle d'astronomie had become a centre of the RASC in 1942 but retained its French name. In 1946, Nadeau was awarded the Chant Medal. In Montreal, the centre established in 1918, notwithstanding its francophone first president, was predominantly anglophone. In 1930, of some 115 members, only about 15 were francophones. A separate Franch-speaking group was formed and admitted as a centre in 1947. The Centre d'astronomie de Montréal has continued its separate existence but has never been as large as the Montreal Centre, indicating that in Quebec popular interest in astronomy – at least in joining an organization – differs from that in the rest of Canada.

Astronomy was not popularized in English Canada early in the century. Periodicals and newspapers offered little coverage. For example, Eddington's 1919 announcement that his observations had confirmed Einstein's general theory of relativity elicited almost no media response.[38] The RASC's centres provided some activities for the general public, while both federal observatories and the David Dunlap opened their doors to the public for viewing.

Chant was particularly concerned about the low level of awareness of astronomy. In 1926, he prepared a set of 50 slides with an accompanying booklet on the solar system especially for groups such as scouts; the University Extension at Toronto prepared the sets. It was the school system, he recognized, that had to be enlisted, and after discussions with the chief inspector and the chairman of the Toronto Board of Education, he agreed to prepare a 100-slide set with a sixty-page booklet for use in the schools. The officials were enthusiastic and ordered sixteen sets, and Chant talked with principals to encourage their use. Unfortunately, as he wrote to Lick's R.G. Aitken, who had supplied a number of the slides, 'But the teachers are themselves untaught and they do not like additional burdens put on them. They have much mental inertia.'[39] This situation has not changed substantially since.

The new medium of radio was worth exploiting, and W.E. Harper was the first to take advantage of it. In 1928, he initiated a series of short talks on astronomy over CFCT (later CFCF) in Victoria. This series continued until at least 1932, and a number of talks were published in the society's *Journal*. The idea spread quickly: in 1930 H.R. Kingston gave talks on

CJGC in London, in 1931 A.S. Eve and A.V. Douglas spoke on CKAC in Montreal, and in 1932 Chant gave talks for the CBC and CFRB in Toronto. At the same time Chant suggested that a regular astronomical column could be syndicated for newspapers. The RASC responded favourably, and Southam Publications agreed to publish fortnightly articles in its papers in Hamilton, Ottawa, Winnipeg, Calgary, Edmonton, and Vancouver. Private papers in Victoria and Halifax also joined in. The series lasted from 1929 to 1931, and virtually every Canadian astronomer provided copy. During the Depression, however, Southam dropped the feature in 1932, for economic reasons. No syndicated feature replaced it, though columns did appear in individual papers.

During the nineteenth century, there were so few Canadian astronomers, whose work required only minimal contact with the larger astronomical community, that international connections were of small importance. The few fellows of the RAS living in Canada rarely went to England for meetings. The Royal Society of Canada had facilitated contacts within the country but offered little in the way of international contact. By the end of the century, the North American astronomical community was large enough to sustain a professional organization, and the Astronomical and Astrophysical Society of America was duly formed and held its first regular meeting in 1899. It replaced, for professional astronomers, the AAAS, which had met in Canada three times before 1900 but held little attraction for Canadian astronomers. The new organization provided an excellent opportunity to create contacts with leading American colleagues. Despite continuing links with the RAS, most international contact for Canadians was continental from the turn of the century.

The American group held most of its meetings at the major US astronomical centres, then mostly in the east and midwest, readily accessible by rail from Canada. J.S. Plaskett began fairly regular attendance in 1908, reading papers on the Ottawa coelostat and on his researches on spectrographic camera lenses and slit width. Several of his colleagues attended the early meetings and read papers; Stewart and De Lury were regular attendants. At the Cambridge meeting in 1910, just before the Solar Union conference, Plaskett offered to hold the next meeting in Ottawa. This was accepted, and the society met at the Dominion Observatory in August 1911.[40] Although US attendance was smaller than usual, several notables came, including E.C. Pickering. As we have seen, this meeting offered Plaskett and King an opportunity to press their idea of a large reflecting telescope for Canada. It also was probably the most complete gathering of Canadian astronomers and those of allied sciences in one place for two decades. Plaskett read a paper, and was elected to the council, along with Harper, C.C. Smith, and Stewart.

The next meeting with large Canadian input was that of 1918, when Plaskett, Harper, Motherwell, De Lury, Cannon, Young, and Henroteau all offered communications. Klotz was elected to council, in recognition of his new post as director of the DO. At the 1920 meeting, Klotz became vice-president of the society; fourteen papers were contributed by the government observatory staffs. During the 1920s, younger Canadians began attending, including Pearce, R.M. Petrie, and W.H. Christie. A noticeable fall-off of attendance by DO staff members occurred, coinciding with their general withdrawal from publishing in major journals. The American Astronomical Society, as it was renamed, met a second time in Ottawa in 1929. Again, the American turnout was small, but virtually all the Canadians attended, with papers being read by DO and Geodetic Survey staff members De Lury, French, Hodgson, A.H. Miller, Smith, N.J. Ogilvie, J.P. Henderson, Henroteau, McClenahan, and McDiarmid. Pearce, Beals, A.V. Douglas, and Young provided the astrophysical content. Canadian astronomy's theatrical debut took place at the Harvard meeting in December 1929–January 1930, when the *Harvard College Pinafore* was performed by the Harvard staff and students, including Peter Millman, then a graduate student at Harvard, and Helen Sawyer, soon to marry Canadian-born graduate student Frank Hogg.

For the astronomers at Victoria and the western universities, whose contacts were closer with Lick and Mt Wilson, the Astronomical Society of the Pacific offered a useful organization, but, as C.S. Beals remarked later, although the Victoria group attended the meetings it never felt dependent on its American contacts.[41] More than any other astronomer in Canada, Plaskett was a great believer in international co-operation, as his role in the Solar Union shows. The organizer of the 1910 Pasadena meeting, George Ellery Hale, was the foremost organizer of scientists of the time. During the First World War he instigated the American National Research Council of the National Academy of Science. After the war, he turned to organizing an international astronomical organization that would be more inclusive than the Solar Union. This was to become the IAU, a natural outgrowth of the Solar Union, established by Hale and a committee of the National Academy in 1904. The Solar Union's first meeting was held in St Louis in 1904.

Plaskett was an admirer of the Solar Union,[42] and so he was receptive to Hale's new venture. Writing to Hale in 1919, Plaskett remarked that he wanted Canada to participate in the proposed union and that, properly, the Royal Society of Canada ought to be the adhering body, though it was doing nothing.[43] Hale replied that Canada should join not only the IAU but also his other creation, the International Research Council, as well as the International Geodesic and Geophysical Union.[44] Klotz had been an active participant in the International Seismological Association, but, although Canada had an official connection with that group, there was no scientific society in Canada acting as a national member.

Notwithstanding the Royal Society's inertia, Canada became a member of the IAU and was assessed a financial contribution in 1920. The first formal meeting of the union was held in Rome in 1922, although an organizational meeting was held in Brussels in 1919, the Canadian delegation being Klotz for astronomy and Deville for geophysics. To organize Canadian affairs for the union, a National Committee was established in 1920, consisting of Klotz, Plaskett, Stewart, De Lury, Henroteau, Harper, Chant, Miller, Smith, Young, and L.V. King, a McGill physicist. At the same time, the new International Union of Geodesy and Geophysics was joined by Canada; its National Committee consisted of Klotz, French, and others.[45]

At the 1922 meeting of the IAU, a number of Canadians were first appointed to the union's commissions and made a variety of proposals to them. Klotz was appointed to commissions 3 (Notations), 18 (Wireless Telegraphic Longitudes), and 19 (Latitude Variation); Plaskett to 27 (Variable Stars), 29 (Stellar Classification), and 30 (Radial Velocities); De Lury to 15 (Solar Rotation); Chant to 22 (Shooting Stars); Henroteau to 30 (Radial Velocities); and Stewart to 31 (Commission de l'heure). Among the proposals made were a number by De Lury for solar rotation work, by Plaskett for improvement of stellar classification and radial velocity measurements, and by Young for a third catalogue of spectroscopic binaries, the first two having been published by the Lick Observatory.

Plaskett, Chant, and Daniel Buchanan were present at the 1928 Leiden meeting. The Germans organized a meeting of the Deutsche Astronomische Gesellschaft for a week later than the union assembly, and both Chant and Plaskett attended the meeting at Heidelberg. During this visit Chant saw his first Zeiss planetarium. The next meeting, at Cambridge, Mass, in 1932, had a larger Canadian contingent of eight.

Over the years Canadians took an active part in the union's affairs, as members of commissions and as officers. Both Plaskett and Pearce had been presidents of commission 30 (Radial Velocities), McKellar was a president of 15 (Comets), and R.M. Petrie was the first Canadian to hold the post of vice-president in the union.

The British connection in Canada, revived during the First World War, waned in the aftermath, but for astronomers election as FRAS was still prized. Ottawa staff members elected included Klotz, De Lury, Henderson,

Henroteau, McDiarmid, Smith, and Stewart; Victoria staff included both Plasketts, Petrie, and Beals; before 1931, Chant, Douglas, and Kingston were also elected. As noted earlier, the DO members contributed little or nothing to the society's publications, but the Victoria staff were more involved. The greatest honour came in 1930, when J.S. Plaskett was awarded the coveted gold medal and was named Darwin Lecturer for the year. In fact, Plaskett was probably the most honoured Canadian astronomer, garnering the Rumford Prize of the American Academy of Arts and Sciences in the same year and the Bruce Medal of the Astronomical Society of the Pacific in 1932. He was also one of the few Canadians chosen FRS.

The AAAS, in which Canadians took little part normally, met in Toronto for the second time in 1921. A joint session was arranged with the RASC, and eleven Canadian astronomers presented papers to the astronomical sessions. The British Association held its fourth meeting in Canada at Toronto in 1924, and Chant arranged a luncheon for astronomers at the University of Toronto's Hart House; afterwards, Eddington spoke on relativity to a large gathering, the talk being broadcast.[46]

Canadians took part also in the Pacific Science Conferences. The first such meeting, held in Honolulu in 1920, brought together scientists from the nations around the Pacific rim. The fifth conference was held in Victoria and Vancouver in June 1933. Plaskett was chairman of the astronomy sessions, attended largely by American astronomers, and both he and Beals read papers. The conference also provided Plaskett an opportunity to show off the observatory to many who had not seen it.

The Second World War naturally curtailed not only astronomical activity but also international contact. The pre-war period saw most of Canada's professionals making attempts to forge international contacts through meeting and conference attendance; after the war, these international contacts, an established part of being a Canadian astronomer, were a sign of full professionalization of the field.

The period 1905–45 was critical for the growth of Canadian astronomy. That government astronomy became fully mature we have seen; however, unless academic support emerged, Canadian astronomy would remain very limited in terms of recruitment, formation of manpower, and areas of research, in short, the necessities of a full-fledged science. This support has to operate on two levels: undergraduate, where students are drawn to astronomy, and graduate, where students are trained for professional astronomy. If we survey the Canadian students who became interested in astronomy as a profession, we find that the first surge – if one can use the word – of undergraduates did not appear until after 1900. When the students began to graduate, they did so almost exclusively at the University of Toronto. This dominance has continued to the present. Specific institutional factors make some universities more attractive than others in certain fields. In the sciences, few Canadian universities offered much scope around 1900. Each of McGill, Toronto, and, to a lesser extent, Queen's was big enough with sufficiently large scientific staffs and adequate laboratory facilities. For Canadian astronomy, however, the personal factor is obviously more important. C.A. Chant's commitment to astronomical education was the decisive factor at Toronto; certainly not decisive were the facilities, until 1935, nor the size of the staff, which numbered but two until the 1930s.

Graduate education in astronomy did require sufficient staff and facilities, and Toronto, with the increase in staff and the opening of the DDO, was the first school to be able to take advantage of the Canadian market for graduate studies in astronomy. That few worked towards the master's degree, and none the PH D, until the early 1950s reminds us that the total numbers of those interested in astronomy as a profession remained very limited until the 1960s; this was as true in the United States as in Canada.

Public interest in astronomy, slight in the nineteenth century, grew steadily in the present century. Canada's great size and dispersed population meant that organized amateur activity was bound to centre upon cities. Further, the number of active amateurs, never large, militated against the formation of purely Canadian specialist groups for meteoritics, lunar and planetary observation, or variable star studies, as in the United States. Instead, the RASC, itself a local organization for Toronto amateurs and professionals, made the transition to national status and was able to satisfy a wide variety of amateur and popular interests. Indeed, the society existed for more than a century, catering to both professional and amateur interests, before a separate professional organization emerged. Chant's organizing abilities were evident in the growth and maintenance of the society; in most local centres of the society, professionals committed to the amateur-professional alliance sustained interest.

The RASC, however, did not professionalize Canadian astronomy. Once Canadian professionals had permanent, scientific employment within the country, and a system of higher education to ensure the flow of new members into the profession, they required means for maintaining their status in the wider scientific community. Their participation in international organizations and conferences increased their sense of professionalism by placing them in regular contact with their foreign peers. Membership in the RASC or even the Royal Society of Canada could not fulfil these critical needs. Foreign contacts opened new possibilities for co-operative ventures and, more important, evaluation of research. This last is absolutely essential for modern science.

The most obvious linkages formed by Canadian astronomers before 1939 were those with American astronomers. The twentieth century saw the slow replacement of traditional astronomy by astrophysics, a field much cultivated by Americans. Proximity, similarity of research programs and equipment, educational system, and language, made them inevitable partners and colleagues. None the less, Canadian astronomy produced its own research, which stood independently alongside American efforts, despite similarities. The professionalization of Canadian astronomy was completed between 1905 and 1940; in fact, by the 1920s, Canadian astronomers were considered peers by their foreign counterparts.

PART FOUR

Canadian Astronomy

SINCE 1945

Dr Helen Hogg on observing platform, 74-inch telescope, DDO, c. 1950,
showing novel method of lowering photographic plate

Visual meteor observers, National Research Council, Metcalfe Road, c. 1950.
At rear are D.W.R. McKinley (left) and P.M. Millman.

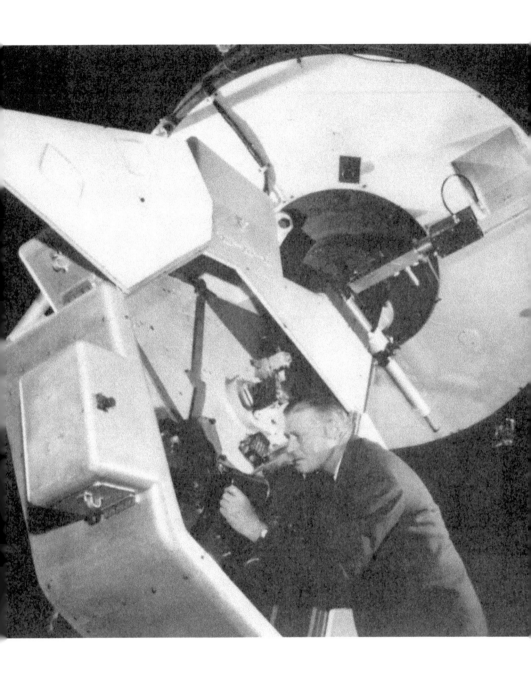

Robert M. Petrie at spectrograph of 72-inch telescope, DAO, 1956

Gerhard Herzberg in his NRC laboratory, 1951

Alice Vibert Douglas at eyepiece of 15-inch telescope, Queen's University, c. 1960

Malcolm Thomson and the mirror transit telescope, DO, 1960s

Adjusting feed horn of the 25.6-m telescope, DRAO, 1960

OPPOSITE

TOP: Scientific staff, DDO, 1964. Left to right: S. van den Bergh, H.S. Hogg, D.A. McRae, R.J. Northcott, D. Fernie, and J.F. Heard (director)

BOTTOM: Third-year astronomy class, University of Toronto, taught by Ruth Northcott (centre rear), c. 1957

The 46-m telescope, Algonquin Radio Observatory

CHAPTER SEVEN

New Centres, New Directions

The Second World War, like the period around 1905, marked a watershed in Canadian astronomy. Until 1945, the Canadian astronomical community was located almost entirely in three institutions – the Dominion Observatory (DO), the Dominion Astrophysical Observatory (DAO), and the Department of Astronomy of the University of Toronto. The post-war era brought decentralization, a multiplicity of institutions, increases in students and universities teaching them, and, perhaps most important, a broad range of new research lines to be investigated. The new field of radio astronomy, born by accident during the war, became an important subject of the National Research Council (NRC), which had hitherto ignored astronomy. The DO, under the guidance of C.S. Beals, quickly expanded its interests and became successful in a number of new areas. In a sense, it succeeded too well: by 1970 it was gone, a victim of government reorganization.

University education, almost the preserve of Toronto, expanded in the great educational explosion of the 1960s and 1970s. By the mid-1970s, two-thirds of Canada's institutions of higher learning offered courses in astronomy, with nearly a dozen providing professional training. Observatories, both optical and radio, sprang up across the country, from six in 1945 to twenty by the end of the 1970s. This was a result of the boom in academic interest in the subject as well as the greater availability of funds from provincial governments and the federal government through the NRC. During the 1950s and 1960s, astronomers nurtured the dream of a major, new observatory, the first since 1935. This resulted in a failure – the Queen Elizabeth II telescope – and a success – the Canada-France-Hawaii telescope.

Of course, expansion was not solely a Canadian phenomenon: it swept the world. Today it is a cliché to mention the scientific awakening brought about by the 1957 launching of the first artificial satellite, but the fervour in US astronomy in the early 1960s had its counterpart north of the border. This incredible expansion in education, particularly in the sciences, was also a concomitant of the post-war 'baby boom.' No one has clearly explained this quickening of tempo. In Canada's case, there was no obvious commitment to a 'space race' with the Soviet Union as in the United States, but the similarity of Canadian and US scientific and educational communities, coupled with the powerful influence of American media, created spill-over. Not surprisingly, this movement created closer ties between the astronomical communities of both countries, with American-trained astronomers increasingly filling Canadian posts and Canadian graduates going south for opportunities in universities, government, and industry. Some idea of the expansion of Canadian astronomy since 1945 can be gauged by the number of professionals: 75 in 1958 and 271 in 1971. Spill-over also appeared in the popular acceptance of astronomy. The RASC has grown to nearly 3,500 members across the country, and planetariums were built in most major centres.

By the end of the Second World War, the Victoria and Toronto observatories were well entrenched as important centres of astrophysical research, while the DO, despite its steady pursuit of more routine, practical investigations, had faded from international notice. R.M. Stewart remained at his post beyond retirement as part of the war effort, but on his departure in 1946, the Department of Mines and Resources evidently decided not to replace him with a staff member. J.A. Pearce was approached but declined to go to Ottawa. The head of the Surveys and Engineering Branch then called on C.S. Beals. Beals, who had just completed a program of obtaining spectrograms of P Cygni stars, was willing to try something new and went, in an acting capacity, in the autumn of 1946. Despite some pressure from the Public Service Commission for him to return to Victoria, he remained and was appointed dominion astronomer and director in 1947.

Why an astrophysicist was sent to an institution devoted to practical work was not made clear, but Beals believed that he was considered an innovator by the ministry, which wanted to shake up the staid observatory.[1] Among his first acts as head were to increase the number of younger members and to stimulate the others to publish.[2] This stimulation made a significant difference in the observatory's output; apart from the studies of Henroteau, little astronomy had appeared in the observatory's *Publications* before Beals's arrival, nor had many geophysics papers appeared. Between the late 1940s and 1970, nearly 150 geophysical papers appeared in the *Publications*. The astronomers tended to publish in journals, and a reprint series, the *Contributions*, was initiated. By 1970, some 300 'contributions' had been produced by the geophysics and astronomy staff members.

One area of research relatively undeveloped in Canada, and one in which the observatory would heavily involve itself, was meteor astronomy. The study of meteors and meteorites was not a new subject in Canada; the *Journal* of the RASC had carried articles on these topics for years, and Chant had devoted almost an entire issue to the spectacular train of meteors seen in Ontario in 1913.[3] The virtual founder of the field in Canada, however, was Peter M. Millman. As a student assistant at Victoria he had studied spectroscopic binaries like most of the staff but, while at Harvard studying for a doctorate, became interested in meteor photography and spectra. On completion of his degree in 1932, he remained at Harvard for a year for further research. He was appointed to Toronto in 1933. At the DDO he and student assistant K.O. Wright photographed meteors and obtained a few spectra. He was able to obtain the co-operation of the DO's Miriam Burland and Malcolm Thomson. By 1941, the group had amassed some 20,000 observations before the work was disrupted by the war.

In March 1946, Millman, then a member of the Research and Development Division of the Royal Canadian Air Force (RCAF), met in Cambridge, Mass, with F. Whipple of Harvard, Z. Kopal of the Massachusetts Institute of Technology (MIT), and W. Bleick and R.J. Seeger of the US Navy to plan a co-operative program of meteor astronomy.[4] MIT would take charge of studying the altitudes and velocities of meteors, while further spectroscopic work would centre on Ottawa. With R.M. Stewart's concurrence, Millman committed the observatory to this venture; he, himself, had just been selected to head the observatory's Stellar Physics Division. The ideal instrument for direct meteor photography was a new wide-angle Schmidt camera designed by J. Baker in the United States. This so-called Super Schmidt was developed by the Perkin-Elmer Corporation and was ready by 1951. Four were to be operated by Harvard on behalf of the US Navy and Air Force, and two were earmarked for the DO's program. These were moved to Meanook and Newbrook, Alberta, and went into regular operation in 1952.[5]

The study of meteors, along with solar radio observation, brought the NRC into astronomy. The NRC had been founded during the First World War as the Honorary Advisory Council for Scientific and Industrial Research to co-ordinate scientific war work but, by the 1930s, had its own laboratories and research staff in Ottawa.[6] Apart from giving grants and fellowships for astronomy, the NRC took no active interest in the subject. After the Second World War, when the Radio and Electrical Engineering

Division – responsible for wartime radar research – was on the outlook for new research projects, radio and radar astronomy seemed obvious fields to cultivate.

The first radar echoes of meteors were recorded in Britain by Hey, Parsons, and Stewart in 1946, although some exploratory work had been done two years earlier. The DO and the NRC linked up to pursue this subject. The NRC participant was D.W.R. McKinley, who provided transmitters of 50–400 kW, which broadcast at 32.5 MHz; the sets were designed and built by Radio and Electrical Engineering Division and erected at the Metcalfe Road Field Testing Station on the south side of Ottawa, at Carleton Place, and at Arnprior. By 1948, Millman and McKinley, filming the echoes on cathode tubes, had some 100,000 observations, mostly for meteors at altitudes of 80–100 km.[7] The joint meteor work advanced along several fronts: McKinley continued with radar studies and was able to ascertain velocities of large numbers of meteors by radar by 1951;[8] Millman continued his interest in meteor spectra. Both had found that meteor trains persist not only visually but also maintain 'visibility' to radar for some seconds. This, Millman assumed, was caused by ionization created in the upper atmosphere as the meteor lost material; recombination and de-excitation of ions and neutral atoms in the atmosphere were the source of the train.[9]

Millman's spectrographic studies resulted in some 120 spectra by 1952, and half of them had been obtained in Canada. In 1955, Millman left the observatory for the Upper Atmosphere Research Section of the NRC and was succeeded by J.L. Locke, who placed Ian Halliday in charge of meteoric research. The major post-war event for meteor research was the International Geophysical Year (1957–8) program of observations, in which Canada was a natural participant. The council and observatory co-operated in this venture, the former with radar observations of the back-scatter of echoes at 9.2 m, while the latter employed organized teams of visual observers, both amateur and professional. This program was co-ordinated by Millman for the NRC and Miriam Burland for the observatory. The chief venue of the program was the Springhill Meteor Observatory, built by the NRC south of Ottawa in 1956. At Springhill, visual observations by observers in heated 'coffins' complemented radar, spectrographic, and, eventually, videotaped recording of meteors. Completion of the IGY program in 1958 was essentially the end of the most active visual work on meteors in Canada; reduction and analysis of data followed.

The DO entered another branch of meteor research, the study of meteoritic impact structures, after the discovery of the New Quebec Crater.

This crater, in the Ungava region of Quebec, although photographed by the US Air Force and the RCAF in the 1940s, was not noticed until Fred Chubb, a prospector, contacted V.B. Meen of the Royal Ontario Museum in Toronto in 1950. Meen, Chubb, and a photographer, supported by funds from the Toronto *Globe and Mail* and other donors, explored the crater briefly in July of that year. Some 3,000 metres in diameter, the crater suggested meteoritic origin.[10] A second expedition was mounted the following summer by the museum and the National Geographic Society. C.S. Beals, by then dominion astronomer, became personally interested in the problem – a major shift from his earlier research – and initiated a survey of maps and of thousands of aerial photographs of the Canadian Shield for evidence of other possible impact craters. A number of likely candidates presented themselves, including the Manicouagan feature, many times larger than the New Quebec Crater. The observatory sent out field parties to survey the gravity profile, to test the residual magnetism of the rocks, and to study the pattern of rock fracturing.[11] This work has continued to date, with much success. Undertaken by large teams, it was a fine example of the collaboration between astronomers and geophysicists fostered by Beals. Between the wars there had been little real contact between the two.

In 1960, the NRC created an Associate Committee on Meteorites, which recognized that the actual recovery of meteoritic material could rely upon more than chance with an automatic patrol camera system.[12] The DO took up the idea; site testing for a network began in 1966, the first station of the Meteorite Observation and Recovery Project (MORP) was in place two years later, and, by 1971, a dozen stations across the three prairie provinces were operational. This area provides relatively level terrain for recovery and largely unobstructed horizons. That the system worked was proved by the recovery, in 1977, of the Innisfree meteorite in Alberta. With this shift in interests, the older visual station of the NRC at Springhill declined in use while the Meanook and Newbrook camera stations were closed in 1970.

Meteor astronomy, although little populated by professionals internationally, has developed into a strongly Canadian branch of astronomy.[13] The DO and NRC groups, by combining their expertise in visual, radar, instrumental, and geophysical aspects, have pioneered many important advances. Besides those already noted, Canadians were the first or among the first to employ panchromatic and infra-red emulsions, transmission gratings, rotating shutters, and airborne observations. Ian Halliday's identification of the auroral green line in meteor spectra in 1958 was a significant step in spectral analysis of meteor-atmosphere interactions. Meteor astronomy, being relatively inexpensive and not requiring high-quality optical and radio sites for observation, was well suited to the Canadian environment and scientific manpower.

In radio astronomy the pioneering work of Karl Jansky and Grote Reber was essentially ignored by the scientific community during the 1930s and 1940s, and only accidentally, through radar research, did the subject gain any followers during the war. At war's end, the NRC, the radio branch of which was deeply involved in radar work, was in a position similar to US and British agencies in radar research in having few new programs to undertake. Radio astronomy offered an excellent use of talent assembled during the war.

Among the council's radar team was Arthur Covington, a UBC undergraduate before the war who had spent two years studying physics at Berkeley before joining the Radio Branch. Inspired by Reber's work and noting Southwarth's methods and equipment in his observations of solar centimetric radiation in 1945, Covington was interested in undertaking similar studies and suggested a modest project to W.J. Henderson late in that year.[14] The proposal was readily accepted. This was a natural progression, as the Canadian experience with radar was chiefly at centimetre wave-lengths. In July 1946, a 4-foot paraboloid, constructed from surplus army radar parts, was erected at the Metcalfe Road station. The radar of this type operated at 2,800 MHz, and Covington built a Dicke switch, invented a year earlier but just described, to create a more sensitive radiometer.

After tests with the instrument at 3,000 MHz on sky noise,[15] observations of the solar disc were begun on 26 July 1946, although no method of calibration had been determined. Covington initiated regular, daily observations in February 1947; these continued for more than thirty years. The intention was to measure the radiation, or flux, from the solar disc at 2,800 MHz, but several problems presented themselves. The beam-width of the first paraboloid was some 7°, while the apparent size of the solar disc is only about 30', so that some sky temperature is confused with the direct solar flux.

Covington expected to measure the average flux from the entire disc, but on 26 July a large sunspot group was visible, enhancing the flux. An 'average' solar flux could not be recorded unless the sunspot enhancement were accounted for; this was possible only after the group rotated to the solar limb. Since the small paraboloid had a beam-width far larger than the solar disc, which, when compared with the sunspot group, had an angular size much larger, the flux density due to the sunspots could not be ascertained. The way around this difficulty was by means of observation of the partial solar eclipse of 23 November, which was, by happenstance, observed visually at the DO. With the moon's limb occulting the sunspots, one could determine the difference in radio 'temperature' of the different parts of the sun. The surprising result was that the solar disc seemed to have a temperature of 56,000 K, not the visually observed and theoretically accepted value of 6,000 K, while the sunspots seemed to have a temperature of 1,500,000 K, when they were generally assumed to be cooler than the photosphere.[16] This result was confirmed elsewhere, and theoreticians soon showed that such radiation arose in the sun's atmospheric layers which, because of their lower densities, attain much higher temperatures.

In 1947, the radio telescope was moved to Goth Hill in South Gloucester, south of Ottawa, and daily observations commenced on 14 February. Covington, believing that geomagnetic activity could be correlated with solar radio noise, thought that a daily record of centimetric radiation would be valuable comparative data if one assumed that shorter-wavelength solar radiation, primarily ultraviolet and X-rays, varied at the same rate. These shorter wavelengths are almost entirely absorbed by the earth's atmosphere and give rise to ionization, while the radio waves are little attenuated. The first series, known as series A, lasted until 1955, when corrections were made after many tests by Covington, N.W. Broten, W.J. Medd, G.A. Harvey, and others, particularly in Japan.

Among these tests was the important study of the sky temperature, which would affect the flux measurements, since the beam-width of the telescope was larger than the solar disc. Initially this was checked by pointing the paraboloid upwards. In 1948, Covington and Medd experimented with a small yagi antenna to measure solar flux at 1.5 m and evaluate the short bursts of flux.[17] To increase resolution, a new instrument was designed and built in 1952; it consisted of a 150-foot slotted waveguide in a trough with movable flaps acting as a horn. This was capable of a beam-width of 8' in an east-west direction. Further improvements transformed the instrument into a 600-foot compound interferometer with a beam-width of 1.2', which, in 1958, was excellent resolution. With this telescope, the NRC group confirmed D.F. Martyn's 1946 prediction that quiet sun radiation arises in the corona, thereby giving a flux pattern of a bright radio ring around the sun.

One other important result of the solar patrol was Covington and Broten's 1954 study of the so-called slowly varying component of solar radio flux. By comparing their data with spectroheliograms supplied by the McMath-Hulbert Observatory in Michigan, they showed that this flux was connected with the bright regions of hydrogen and ionized calcium on the solar disk.[18] During the same year, a 10-foot paraboloid was added for studying solar radio bursts.

Up to that point, apart from analytical work by Ralph Williamson at Toronto, radio astronomy in Canada was essentially limited to the NRC group. C.S. Beals had hoped to involve the DO in radio astronomy as early as 1947 but was dissuaded, his chief opposition coming from Frank Davies of the Defence Research Board.[19] Ironically, the board entered limited radio astronomy work before the observatory finally initiated its own program. Beals made further representations in the mid-1950s; establishment of a radio observatory as part of the observatory's programs was approved in 1957, and site-testing was begun. The two best sites to emerge were White Lake, near Penticton, BC, and Lake Traverse, in Algonquin Provincial Park, in Ontario. Beals was able to report to the American Astronomical Society's meeting in 1959[20] that a 25.6-m paraboloid was being constructed in Penticton for use in the decametric range and that J.L. Locke of the Stellar Physics Division of the observatory would be first director of the Dominion Radio Astrophysical Observatory (DRAO). C.H. Costain, and J.A. Galt – a DO staff member who had spent fifteen months at Jodrell Bank in England – were to be the first permanent staff members.

The radio telescope was designed chiefly for use at about 1,420 MHz, the radio spectral line of neutral hydrogen, the most important indicator of galactic structure. The observatory opened officially in 1960, and the staff began studying the distribution of neutral hydrogen in the galaxy.[21] This form of research was under way in many radio observatories around the world, but there had been little study of long-wavelength, low-frequency radiation from both galactic and extragalactic sources, such as the quasars which were discovered in 1960. Solar and man-made radiation renders low-frequency cosmic noise very difficult to detect, but Costain pressed for such research at Penticton. Ottawa agreed, and in 1962 work began on an interferometric array which would operate at 22.25 MHz; this array consisted of 624 full-wave dipoles mounted on wooden poles in a T-formation. Galactic mapping commenced in 1964. Even lower-frequency emissions are detectable only when the solar flux is relatively low, and since the International Year of the Quiet Sun was due to commence soon, a 10-MHz array, similar in size and construction to the 22-MHz interferometer, was ready in the spring of 1965. At the same time a 1.8-m paraboloid was installed as a co-operative venture with the NRC solar patrol and was identical to the instrument in use at Lake Traverse but operating at 2,700 MHz. In 1962, on Locke's return to the DO, Galt became director.

While site-testing was taking place for the location of the DRAO, it became clear to the NRC radio astronomers that the Goth Hill location was too radio noisy for continued use, and consequently they were able to obtain the Lake Traverse site in 1959. Because Toronto and Queen's were developing radio astronomy programs, the radio astronomers decided that the NRC site should become a national radio astronomy observatory, similar to the National Radio Astronomy Observatory in West Virginia, except that it would be operated by a government agency instead of as a university consortium. The solar patrol was the first phase of the Algonquin Radio Observatory (ARO), and a new 6-foot paraboloid to continue the 2,800-MHz observations was installed in 1960 and employed simultaneously with the Goth Hill instrument until 1962. By 1965 it was working in tandem with the Penticton instrument. In 1960, the NRC group began planning a multiple-array radio telescope,[22] which, when placed in service in 1966, had thirty-two, 10-foot paraboloids mounted on a 700-foot base. Operating at 2,800 MHz, the telescope initially had a resolution to 30' but, because of financial and staff limitations, only the basic array has been operated continuously. This capability allowed for the flux density of specific parts of the sun to be measured. A 10-foot paraboloid and a horn telescope rounded out the solar astronomy apparatus at the ARO.

The interest in other radio objects led to a desire for a large, steerable antenna, and planning began in the late 1950s for an NRC paraboloid comparable in size to the Penticton telescope then contemplated. A larger, 46-m dish emerged as a choice, and work began in 1959, but seven years elapsed before the new instrument went into service at Lake Traverse. It has similarities to the 140-foot paraboloid of the (US) National Radio Astronomical Observatory, completed a year earlier; the West Virginia telescope, designed with an equatorial mounting, created many engineering problems. The Algonquin telescope was designed by the British firm of Freeman, Fox and Partners, which had designed the 210-foot altazimuth telescope at Parkes, NSW, Australia. The smaller Canadian version, built by Dominion Bridge of Montreal, is capable of working to wavelengths of 2 cm. By 1967, Canada possessed one of the world's largest and most accurate radio telescopes for galactic and extragalactic research.

Almost immediately after completion of the 46-m paraboloid, a co-operative operation by the NRC group, the DRAO, Queen's, and Toronto gained international notice through successful completion of the first Very Long Base Interferometry experiment. Because the resolution of an object in the sky depends largely on the size of the reception device – lens, mirror, or radio antenna – and since radio wavelengths are considerably longer than optical ones, very large antennae are required for reasonable separation of objects. In the mid-1960s, the most exciting objects in the radio sky were the quasars, radio bright, but distant objects with very small apparent diameters. As the best radio telescopes were incapable of resolutions much better than minutes of arc, better separation required either monster antennae, which were out of the question, or an interferometer. Based upon its optical counterpart used by Michelson early in the century, the radio interferometer, dating in usable form from the late 1940s, brought the radio signals of separated antennae together; the signals interfere with one another, and analysis of the signals yields the angular separation.

Over the years the elements of radio interferometers were separated more and more, first by cables, later by microwave link. Continent-wide bases were impossible until it occurred to a number of groups that the signals could be recorded at different sites if recorded in conjunction with the output of an oscillator controlled by an atomic clock. The recordings could then be analysed together by having a computer search for the interference pattern. Australian, British, and US groups all attempted this in 1967, but the Canadian team, led by J.L. Yen of Toronto, with Broten, Locke, T.H. Legg, C.W. McLeish, and R.S. Richards of the NRC, J.A. Galt of Penticton, H.P. Gush of Toronto, and R.N. Chisholm of Queen's, managed it first.[23] The baseline of 3,074 km – the distance between the two antennae, the 46 m at Algonquin and the 25.6 m at Penticton – was the longest ever attempted, and videotape was employed as the recording medium, as it would accept the suitable bandwidth. The objects chosen were the quasars 3C273 and 3C345; at a frequency of 448 MHz, the resolution was an astonishing 0.02', and the location and size of the quasars in the radio region could be established.

The success of the experiment was reported to an Ottawa meeting of the International Scientific Radio Union in May; four years later the team, along with American workers who soon managed the same feat, was to garner the Rumford Premium of the American Academy of Arts and Sciences. The 46-m paraboloid has since been applied to a variety of radio studies of our galaxy and other galaxies and quasars; in the mid-1970s, a rapidly growing study of interstellar molecules attracted the attention of radio astronomers. Optical spectrographic studies of the interstellar medium, earlier studied at Victoria, were again the object of Canadian attention. In 1978, the NRC team of Broten, L. Avery, J. MacLeod, and T. Oka, with H. Kroto of Sussex, identified the microwave line of the heaviest known molecules in space, HC_7N and HC_9N.[24] These successes led to a proposal to resurface the antenna so that it could operate at millimetric wavelengths, enhancing its microwave capabilities.

Galactic surveys at Penticton were enhanced by the construction of a super-synthesis radio telescope at the end of the 1960s.[25] Aperture synthesis, a technique pioneered by Ryle and others in Britain in the late 1950s, allows the radio astronomer to have the equivalent of a telescope of large aperture with two or more smaller antennae that are movable. The separation of the elements and the motion of the earth carrying the 'beams' of the telescopes to different positions allow the various signals to be synthesized into a coherent whole. The Penticton instrument, consisting of two 8.5-m paraboloids on a 600-m east-west track, can synthesize images equivalent to that of a 900-m paraboloid with about a 1' resolution for work at 1,420 MHz.

The advent of radio astronomy and the expansion of meteor research enriched Canadian astronomy, yet more traditional aspects were not ignored. Post-war astronomers took a variety of new initiatives. At the DO, the Positional Astronomy Division, in charge of meridian work and the time service, was under the direction of W.S. McClenahan until 1957. Work continued with the meridian circle along established lines until it was retired from service in 1962. Well before then, however, Beals was seeking new ideas. One project that no one had attempted was to build a mirror-transit telescope, proposed in 1947 by British astronomer Robert d'E. Atkinson in an article in the *Monthly Notices*. By employing mirrors instead of the usual objective lens and through automation of its operations, the mirror-transit telescope could eliminate personal equation in measurement. Beals and the divisional staff, meeting in May 1953, agreed to proceed. Atkinson was brought over for consultations, and an experimental model was built. Unfortunately, there were many difficulties in design and operation; after six years of effort, Beals realized that the cost to perfect the instrument was too great, and the project was dropped in 1959.

More successful was the adoption of the photographic zenith tube (PZT), invented early in the century and used with great precision at the US Naval Observatory, where it made the transit instrument obsolete for the recording of star transits. McClenahan was able to consult with American astronomers on a similar instrument for Ottawa. Completed in 1951, the Ottawa PZT became, in the following January, the observatory's instrument for time determination. Improvements were made by M.M. Thomson and others, and by 1965 a second instrument was ordered, so that the older PZT could be moved in 1967 to Priddis, Alta, where observing conditions were superior. The newer instrument was eventually relocated outside Ottawa.

Solar astronomy at the DO was the domain of J.L. Locke in the early 1950s. Since solar rotation study had ended with the retirement of Ralph De Lury, new instrumentation would be required to embark on different lines of research. A grating spectrograph was added to the solar telescope, and a Lyot filter was fitted to the 15-inch telescope for studies of the sun in Hα, particularly for the photography of flares. This was automated. During the IGY (1957–8), the 15-inch was employed as a flare patrol camera, and the program continued into the early 1960s. By then it became clear that the site at the Central Experimental Farm was no longer suitable for solar photography, and site-testing outside the city began. Shirley's Bay, some 20 km west of city centre on the Ottawa River, was chosen, and the department earmarked funds for a new solar telescope. Completed in 1970, the Ottawa River Solar Observatory[26] features several small refractors mounted on a spar, the major component being a 25-cm photo-heliograph capable of observations of the whole solar disc in Hα, photospheric photography in various wavelengths, and small-region Hα photography of the chromosphere. The instrument is also capable of cinematography, which has enabled Victor Gaizauskas of the NRC to undertake detailed studies of rapid solar atmospheric changes.

In June 1962, Beals retired as dominion astronomer and director of the Dominion Observatories Branch of the Department of Mines and Technical Surveys. This signalled a reorganization of astronomy within the department. The position of dominion astronomer went to R.M. Petrie of Victoria, in recognition of his position as senior astronomer of the service, but with the title severed from the Ottawa directorship. J.H. Hodgson of the Ottawa staff was elevated to director of the Observatories Branch, with direction of the Ottawa, Victoria, and Penticton observatories, magnetic observatories, seismological stations, and the meteor stations, all operating as separate units. This arrangement lasted only six years, until the larger reorganization of 1970, the only interim change being the restructuring of the ministry in 1966 into the Department of Energy, Mines and Resources (EMR).

After the war there was no immediate expansion at Victoria. In 1951 Pearce took early retirement and was succeeded by Petrie, who held the post until his death in 1966. Petrie was chiefly involved in the new B-star program initiated with Pearce in 1942, a project aimed at obtaining spectra of all B-type stars earlier than B7 and to 8.6 magnitude. This program was not complete until 1962, when a catalogue based upon some 2,300 spectrograms appeared.[27] One of Petrie's interests was the establishment of a correlation between the strength of the Hα lines in the spectra and the absolute magnitudes – and, therefore, the distances – of the B-type stars. His 1953 relationship was some 0.7 of a magnitude less than the corresponding values of the Morgan-Keenan classification then in use.[28] The data did, however, reconfirm the Oort hypothesis of differential rotation of the galaxy and showed, with an analysis of the Ca II lines, that interstellar gas is uniformly distributed out to some 3,000 parsecs. Petrie was able also to show that B stars have residual random motions apart from their normal galactic rotation, amounting to an average velocity of + 13 km/s. The final analysis of the project was incomplete on his death, being published two years later by his widow, J.K. (McDonald) Petrie.[29]

Work at the DAO, largely spectroscopic binary and radial velocity research before the war, broadened further after 1945. Anne B. Underhill moved to the observatory from UBC in 1946; although she and William Petrie had studied the Stark effect in O-type stars during the war, she continued research on hot stars, particularly theoretical studies of the structure of early-type stars. She remained until 1962 when she accepted professorship at Utrecht. Other newcomers in the 1950s included E.H. Richardson, a specialist in molecular spectra, who became one of the chief instrument designers at the observatory, and Graham Odgers, an Australian theoretician. Both would become key personnel in the Queen Elizabeth II telescope project initiated by Petrie.

By the mid-1950s, with staff approaching a dozen permanent members plus a number of visitors and summer assistants, the pressure on the 72-inch telescope was tremendous, even though it had always been pushed to near capacity. Richardson noted in 1968 that during the fifty years since it began operation the telescope had been employed in obtaining nearly 66,000 spectrograms at the cassegrain focus.[30] Discussions aimed at obtaining an additional telescope began in 1950, and Petrie and the staff designed an instrument with greater spectrographic flexibility. They settled on a 48-inch reflector of 192-inch focal length, but with a coudé focus, then used only for a few large reflectors. Progress was slow, despite Beals's advocacy in Ottawa: the expenditure was agreed to in 1954, but a contract was not let to Grubb-Parsons until 1957. The completed instrument was installed during the winter of 1961–2. Richardson and the optical staff worked to improve the coudé operation and, with the devising of an optical image slicer together with highly reflective coatings for the mirrors, were able to produce a coudé spectrograph as fast as that of the 200-inch telescope at Mt Palomar.[31]

On Petrie's death in 1966, K.O. Wright became director, and in the decade of his leadership the staff, and the number of lines of research, continued to increase. A 16-inch telescope was added, and a new optical shop with testing tower was built in 1973. In the following year, a new Cer-Vit mirror replaced the old 72-inch plate glass disc.[32] Although the variety of research increased with new equipment and personnel, the traditional field of binary stars was carried on by A.H. Batten, who joined the staff in 1959; in 1967 he published a new catalogue of spectroscopic binaries,[33] a project undertaken since 1905 at the Lick Observatory. Batten managed to do what J.S. Plaskett had hoped to accomplish some forty years earlier.

Finally, one research group not directly connected with government astronomy but making important contributions to the science was the NRC spectroscopy group, created in Ottawa in 1948 with Gerhard Herzberg as head, along with his Saskatchewan colleague, A.E. Douglas. This group attracted an excellent staff and gained international renown for laboratory astrophysics. Among the studies the group performed was the first laboratory identification of anomalous hydrogen bands that optical astronomers discovered in the spectra of Jupiter, Uranus, and Neptune.

By the late 1960s, the number of astronomers in Canada was increasing, the tempo of graduate instruction quickening, and the observing facilities becoming taxed. No major new institution had been created in Canada for more than thirty years, and several astronomers, particularly R.M. Petrie, began thinking seriously about a major installation. This feeling was no doubt stimulated by the rapid increase in large instruments in the United States during the decade. An innovation in American astronomy had an effect on some Canadians: the creation of a national observatory at Kitt Peak in Arizona, funded by the National Science Foundation but operated by a consortium of universities, the Association of Universities for Research in Astronomy (AURA). Although AURA was formed in the mid-1950s, the Kitt Peak National Observatory became a premier astronomical institution only in the 1960s. The Canadian government, unlike its American counterpart, had been heavily involved directly in astronomy for many years, and so it was not surprising that Petrie, head of a government observatory, should emerge as the leader of a movement for a new institution.

A meeting of the Canadian National Committee of the IAU at Victoria in October 1962 was devoted to 'The Future of Optical Astronomy in Canada,'[34] and Petrie urged erection of a telescope of 144-inch aperture. As he noted the following year, 'Our favourable position has now declined and we are in danger of being relegated to a secondary position because we have not planned for a more powerful telescope during the past thirty years.'[35] Kitt Peak was an obvious example to emulate, because such an observatory could service the growing university population as well as government scientists. The National Committee backed the proposal, and a subcommittee began site-testing in 1963 under the direction of Graham Odgers of the DAO. He had already discovered that the region around Penticton, BC, was at least as good as Victoria – but without the urban growth – and in that year he found the Mt Kobau site, near Osoyoos, hard by the US border.

Looking forward to the Canadian centennial celebrations four years later, astronomers began referring to the project as the 'Centennial' or 'Confederation' telescope. The government, requiring some token to commemorate the royal visit of 1964, seized upon the telescope project, and Prime Minister Pearson announced on 12 October 1964 that the telescope would be named in honour of the queen; the project thereafter was dubbed the QE II telescope. While EMR had already provided some funds for the project, such an announcement must have seemed to be an irreversible commitment. Such was not to be.

The planning crystallized in 1965. Petrie's ideas were elaborated upon; a 150-inch reflector was to have a quartz primary and could operate in three modes, an f/30 coudé focus for spectroscopy, f/8 and f/15 cassegrain foci for photometry and photography, and an f/2 prime focus for photography and spectroscopy. After Petrie and Odgers conferred with American colleagues, the blank was ordered from Corning in September for 1967 delivery, and R.E. Dancey was named chief optician. Two major consultant contracts were let: the overall design of the observatory to A.B. Sanderson and Co of Victoria, and the design of the telescope, controls, and mirror-support system to Dilworth, Secord and Meagher of Toronto. Just as the project went into high gear, Petrie died. Odgers stepped in, and along with G. Brealey, E.H. Richardson, and D.H. Andrews, formed the design team. Sanderson's report for the Department of Public Works, which would contract out construction, was ready by the end of March 1967.[36]

This document is remarkable for its breath-taking vision and its almost total lack of political acumen. Petrie, in 1962, suggested a large telescope; the Sanderson report offered nothing less than a Canadian Kitt Peak. Indeed, the title of the report concerns the 'Mt Kobau National Observatory,' not the QE II telescope. Virtually all of government astronomy was to be concentrated on Mt Kobau. This necessitated a restructuring of government astronomy. A National Astronomical Institute was to be created in Vancouver, headed by the dominion astronomer, with scientific divisions for astrophysics, positional work, photometry, solar physics, meteor physics, radio astronomy, time service, and space research. The institute would control six facilities: Mt Kobau, the DAO, the DRAO, the PZT instrument at Calgary, the PZT and a solar flare station at Ottawa, and a meteor station in Saskatoon. The institute's staff would number some 125, with 14 others stationed in Calgary, Saskatoon, and Ottawa.

Even more surprising were the plans for the physical facilities and instrumentation. For astrophysics, the QE II telescope would be joined on the mountain by general-purpose reflectors of 60–80-inch and 48-inch ranges, a 40-inch photometric instrument, and a similar telescope of 16-inch aperture. The last one was already in place for site-testing. Positional astronomers would obtain a 60–80-inch astrometric reflector, and the mirror-transit telescope, which had yet to be perfected, would be moved to Mt Kobau along with the time and frequency laboratory. The Meanook and Newbrook meteor-observing equipment would be moved to the observatory. Alongside these would be a new solar tower and atmospheric research laboratory. A suggested saving was to remove the 72-inch and 48-inch telescopes from Victoria to the new site, thus closing the older observatory.

The costs bore little resemblance to the original plans. The report estimated the cost of the QE II telescope at $12.6 million; the rest of the astrophysical instrumentation would add $2.2 million. Positional astronomy would cost $1.9 million, solar and meteoric astronomy another $1.4 million. The other facilities on the mountain, the 'village,' residence, and visitors' centre, would add another $1.8 million. The Vancouver headquarters and optical shop would take $2.5 million. Altogether, the inflated project would cost some $22 million. The report appeared in a year of budget restraint, and, although the mirror blank for the QE II telescope was delivered late in the year, lack of funds had delayed the construction of the optical shop. This was seen as an irritant; it was actually a harbinger of the collapse of the project.

The Mt Kobau project was to be one of the first objects of scrutiny for the new science policy machinery created by the Pearson government. Science policy, in the sense of setting goals and priorities for Canadian science, had been a largely ad hoc process until the 1960s.[36] In theory, governmental scientific programs were reviewed by the Privy Council Committee on Scientific and Industrial Research, created during the First World War. However, this group – never a formal cabinet committee – met only sporadically. In its report in 1963, the Royal Commission on Government Organization, headed by J.G. Glassco, noted that the Privy Council committee had failed to meet altogether between 1950 and 1958.

As we have seen, most of the policy positions on astronomy taken by government were the outcome of direct communications between astronomers, societies, deputy ministers, and ministers. The early triumphs were effected by direct lobbying by King and Plaskett through men like W.W. Corry, Clifford Sifton, and W.J. Roche. The QE II project was much the same; the bureaucrat behind the scenes was William E. van Steenburgh (1899–1974).[37] Van Steenburgh had risen through the ranks as a researcher for the Department of Agriculture, and, in 1956, he was appointed chief scientist for research and development for the Department of Mines and Technical Surveys. It was he who had piloted through new astronomical projects such as the DRAO. In 1962, he moved up to deputy minister.

Van Steenburgh pressed the QE II project and seems to have been largely responsible for cabinet's acquiescence. Indeed, this personal approach might have worked but for a shift in science policy. The Glassco Commission had recommended creation of a central body to collect and analyse information concerning government-sponsored science, since many of the decisions on science-related funding ended up with the Treasury Board, which had no competence to evaluate such proposals on scientific merit. Prime Minister Pearson called upon Dr C.J. Mackenzie, wartime president of the NRC, to report on the value of the Glassco recommendations; his report in January 1964 agreed that such a body should be created. In April, the government established the Science Secretariat as a data-gathering and analysis group responsible to the Privy Council Committee, then headed by C.M. Drury. Dr Frank Forward, the first director of the secretariat, thought that the QE II project ought to be reviewed, but van Steenburgh's enlistment of Pearson in the project at the time of the queen's visit forestalled further scrutiny.

When the project report circulated in 1967, events began to move rapidly. The Treasury Board was uneasy about the much larger-scale observatory now envisioned. In the spring, the Carnegie Institute, planning its 200-inch CARSO telescope for Chile, inquired of EMR, which had absorbed Mines and Technical Surveys in 1966, whether it wished to co-sponsor the project. The occasional rumblings of scientists in the NRC and the universities began to break out, particularly among the latter, many of whom saw the Chile telescope as more valuable. To evaluate the conflicting claims and requirements, the Science Secretariat was seen as a possible arbiter, a role beyond its intended function. To that end, it named a study group in the summer of 1968, with recently retired C.S. Beals representing government scientific interests, William Wehlau of Western Ontario for university interests, and Dr D.C. Rose, a retired NRC physicist, as a neutral party.

It became clear in the hearings that the University of Toronto group, the largest academic group in the country, was against Mt Kobau on the scale suggested. It was also discouraged by the climatic conditions on Mt Kobau; longer-term data suggested the site was not nearly so fine for seeing as originally believed. In the end, the study group saw two options available:[39] build a large telescope in Chile, or continue with Mt Kobau project on a reduced scale, by completing the QE II telescope and adding another in the 60–80-inch range, while buying into the CARSO project.

The first option would cost $20 million, the second $25 million, and it leaned towards the second. By 1968, the government had already expended some $4.5 million and expected to disburse an additional $14.3 million.

The Science Secretariat's report to the cabinet, showing a clear split among the astronomers, persuaded the government, now headed by Pierre Trudeau, to cancel everything on economic grounds. The scuttling of the project was announced in an EMR press release on 29 August 1968.[40] This move left a legacy of bitterness and also spelled the end of the more personal approach to science policy decision-making. The desire for a large, modern optical installation did not die and was resurrected at the first opportune moment. In the mean time, however, a major government reorganization of Canadian astronomy changed the nature of the game.

Canadian science has always tended to be dominated by government activity, which has been a centralizing force. In astronomy, there was until recent years a division of labour among government organizations with astronomical interests, first between the departments of the Interior and of Marine and Fisheries, later between Mines and Resources (or Technical Surveys) and the NRC. The powerful centralizing and bureaucratizing attitudes of the Trudeau government, however, virtually ended this division of labour. In 1970, Canadian government astronomy was rationalized under the aegis of the NRC.[41] Both the Glassco Commission and the Science Council of Canada had suggested a reorganization, and the new régime went into effect on 1 April 1970. Since the days of King and Klotz, geophysics, geography, and astronomy had been closely linked in the DO and then with the Geodetic Survey under King's direction. As other specialties emerged, they tended to be taken up by several departments: meteoric studies were undertaken by both the DO and the council; radio astronomy was also pursued by both, along with some efforts by the Defence Research Board before its dismantling; geophysics was advanced by the DO and the Geological Survey. From a purely organizational point of view, it must have seemed simpler to combine these – in a sense – competing institutions.

The primary victim was the DO, which was closed in 1970. The geophysical research section was left with EMR as the Earth Physics Branch. The time service of the observatory was transferred to the Physics Division of the NRC and became the Time and Frequency Section, remaining under Malcolm Thomson's supervision until his retirement in 1972.[42] This was a logical move, since the maintenance of correct time was, by the early 1970s, no longer an astronomical subject but virtually a branch of solid-state physics. With the mirror-transit instrument abandoned and the old meridian circle outmoded, timekeeping in Canada had little connection with traditional practices. Likewise, the PZT instruments could be applied to geophysical research, and so the few workers involved in PZT research were reorganized as the Polar Motion Group of the Gravity and Geodynamics Division in the Earth Physics Branch of EMR. The instruments at Shirley's Bay, near Ottawa, and at Priddis, Alta, have since been maintained by that group.

The integration of the rest of government astronomy into the NRC proceeded in two steps. In 1970, an Astrophysics Branch was created in the Radio and Electrical Engineering Division; J.L. Locke, from the observatory, acted as both associate director of the division and chief of the branch. The latter comprised five sections: the Upper Atmosphere Research Section for meteor work, the Victoria, the Penticton, and the Algonquin observatories, and a Radio Astronomy Section. Gaizauskas's small solar optical team was grouped with the last. In 1975, a further reorganization was instigated, and the Herzberg Institute of Astrophysics, named in honour of the Nobel laureate in the Physics Division, was formed to group all astronomical activities of the NRC under one umbrella. In addition to the original Astrophysics Branch, three sections devoted to laboratory astrophysics in the Physics Division came in. The institute, with Locke as director, was composed of seven sections: the DAO; the DRAO; Astronomy, for radio astronomy, including the ARO, maintained separately from 1975 to 1978; Planetary Sciences; Spectroscopy; Spectroscopy of Larger Molecules; and Space Physics. Consolidation was of value in the Ottawa area, where the majority worked, by easing communication; it meant, for example, one group for meteoric studies rather than two. For the observatories, little was changed: both Victoria and Penticton continued to have resident scientific staffs, while Algonquin Park has never had any. K.O. Wright continued as director at Victoria until his retirement in 1976, when, in a break with tradition, he was replaced by a university man – Sidney van den Bergh of Toronto – rather than someone from the observatory.

The amalgamation of government astronomy under the NRC left in its wake the need for some organizational changes in the profession. The Canadian National Committee of the IAU could no longer be connected with EMR, since that department would cease to be the adhering body to the international organization and would be replaced by the council. In 1970, the National Committee became an Associate Committee of the NRC, one of the many liaison groups between the council, academics, and industrial scientists. Henceforth, Associate Committee members who also belonged to the IAU would constitute the National Committee, effectively forestalling duplication. Meeting in Kingston in March of that year, the committee recognized that the number of astronomers and students of astronomy in Canada had reached a threshold suitable for the launching of a strictly professional society. Although the National Committee had operated for some years in a fashion similar to a society, the Canadian Astronomical Society / Société canadienne d'astronomie (CASCA) was formed in May 1971 and held its first regular meeting in November 1971. It soon attracted nearly 150 members. Its annual meetings, like those of the RASC, have been peripatetic, and, after more than a decade of operation, the society has not led to a general withdrawal of professionals from the older society. Indeed, the abstracts of the papers read at CASCA meetings regularly appear in the RASC's *Journal*, although CASCA publishes its own newsletter, *Cassiopeia*.

With government astronomers united in one organization, and a society to voice the profession's opinions, another push for a major undertaking could be made more successfully. Design expertise and optical grinding machinery, legacies of the QE II project, were available, and, although the government was no more likely to fund a major program in the early 1970s than it had been in the lates 1960s, the possibility of a joint project with France was broached, and the Canada-France-Hawaii (CFH) telescope was born.[43] The creation of the CFH telescope was doubtless the most important event in Canadian astronomy in the 1970s, coming nearly forty years after the last significant optical observatory. While Canada had never participated in so large a co-operative project in astronomy before, there were a number of international models available, especially of joint southern-hemisphere observatories. Planning for the observatory was well under way by 1973, when the tripartite agreement between Canada's NRC, France's Centre national de la recherche scientifique (CNRS), and the University of Hawaii was drawn up; it was signed in the spring of 1974. The CFH Telescope Corporation, which operates the observatory for the three parties, had been incorporated a few months earlier. At the time of signing, the French estimate of cost of the facility was NF91 million – ignoring escalation during construction – and all major costs were to be equally shared by both France and Canada, with the University of Hawaii supplying the land, road, power line, and ancillary buildings. The ratio of funds provided, on a 42.5:42.5:15 basis, was to be used for observational time available to each participant. All parties have membership on the board of direction and scientific advisory board.

The apportioning of contracts reflected the relative expertise of French and Canadian scientists and industry. Most of the studies for the telescope had been undertaken by the French before Canada joined the project, and it was agreed that the French would provide the telescope mechanism, drives, aluminizing tank, and most of the instrumentation. Canadians were responsible for the mirror polishing, controls, dome, laboratory and work-

shop equipment, and some of the instrumentation. Mirror blank technology had proceeded further since the time the QE II blank had been
ordered, so a Cer-Vit blank for a 3.6-m telescope, already obtained by the
French, was used. This was ground and polished in the new facility at
Victoria and shipped to Hawaii, where the French-built mounting, based
upon the Hale Telescope in construction but with removable upper end
and computer-operated drives, had been erected. The observatory opened
in September 1979. The CFH telescope, equipped for primary focus, cassegrain, and coudé operation, was designed not only for photographic, spectroscopic, and photometric studies but also for a new area of infrared
astronomy, which few Canadian astronomers had yet cultivated.

Canadian astronomy has always relied principally upon governments for
funds. In the post-war years the federal government, through the NRC
and EMR, was the chief source of money, and this situation, with the
amalgamation of all government astronomy under the aegis of the NRC,
will continue. University-based astronomy has come to rely upon Ottawa
for any major instrumentation – through the Natural Sciences and Engineering Research Council (NSERC), while salaries and other expendi-
tures are met with provincial grants. Apart from Mrs Dunlap's gift, there
has never been significant private philanthropy for Canadian astronomy,
nor is there likely to be any. Canadian astronomers are aware of the situation, and their planning exercises are always directed towards the federal
government. The NRC Associate Committee on Astronomy produced a
report[44] in April 1974 suggesting priorities for both ground- and space-
based astronomy. In addition to the usual pleas for more funds for research,
a number of capital projects were suggested within a ten-year framework.
Most important, after completion of the CFH telescope, which had been
recently approved, was a 25-m radio telescope operating at millimetric
wavelengths to capitalize on NRC excellence in detecting interstellar molecules. Two optical telescopes were projected, one of 1.5 m for Hawaii,
one of 2 m for the southern hemisphere, along with another radio telescope
for low-frequency work to encourage the research under way in Alberta.
A fifth project was a large telescope for Quebec. To date, only the last
project has materialized (see below), and, while the project had the approval of Canadian astronomers, one can only speculate whether political
considerations moved the lowest-priority item to the top.

The 1974 report discussed space-based research in general terms but, like
other group and individual recommendations of the time, pressed strongly
for creation of a Canadian space agency to co-ordinate and fund space
research. The NRC established a Space Science Coordination Office, but
the real initiatives have continued to be abroad, especially with NASA's

space shuttle program and the announcement by the European Space Agency that it would build a space laboratory, Spacelab, to orbit with the shuttle in the 1980s. Together with NASA's satellite programs and the planning for a large orbiting optical telescope, there were obviously great opportunities for a few Canadian astronomers working on space-based research. An ad hoc committee to collect information on possible Canadian participation in the space shuttle program was struck by the NRC Coordination Office in 1976. The next year, a working group of interested parties reported to the ad hoc committee, which, in turn, reported to the NRC.[45] These reports listed the possible means of Canadian participation in both spacelab and shuttle flights. Up to that time, Canadians from eight universities and the DAO had participated in a number of NASA programs. Upper atmospheric research, a long-time Canadian strength, features most prominently in the ad hoc committee's recommendations.

For astronomers, the most significant space project is the US Space Telescope, but, on a more modest scale, a Canadian 1-metre orbiting telescope was proposed by C. Walker and B. Campbell at UBC in 1978. Discussions with NASA and with Australian space scientists led to a joint project, STARLAB, in 1980. Canadians would be prime contractors for the telescope itself, which would perform both direct photography and spectrography. After considerable effort and expenditure, the Mulroney government abruptly terminated Canada's participation in the project in March 1984, citing lack of funds. This was in the same year that Marc Garneau became the first Canadian astronaut to go into space, widening public interest in space exploration. Indeed, the Conservative government had, several times, committed Canada to greater efforts in research and development, but primarily funded by the private sector. As the Canadian aerospace industry is small, encompassing a few firms such as Spar Aerospace – which developed the Canadarm on the US shuttle – and Bristol Aerospace, not much advance could be expected without government support. Just over a year later, the government anounced that it would participate, to the extent of $800 million, in NASA's proposed space station project, along with Japan and several European nations. Unfortunately, even if this project succeeds, little will result for Canadian astronomers. A space agency, long supported by CASCA, did become a reality, but initial funding came about through the stripping of part of NRC's budget; in 1986, an estimated $20 million would be lost from the council's regular programs. One possible cut, horrifying to radio astronomers, is the proposed closing of Algonquin.

These difficulties came at a time when Canadian astronomers were lobbying for their next major installation, the Canadian Long-Baseline Array.

The radio astronomy committee of CASCA suggested the project in 1979, capitalizing upon Canadian experience with Very Long Baseline interferometry, beginning in 1967 but continuing with joint efforts with other nations. The American Very Large Array was just taking shape in New Mexico, and a Canadian version, with eight 25-metre paraboloids located across the country from British Columbia to Newfoundland, could be created for some $23 million. A 1981 feasibility study and subsequent discussion by CASCA have kept the project alive, but, given the present political climate, little action is likely in the near future.

The future of ground-based optical astronomy and of radio astronomy is severely limited, not only because of deteriorating sites in Canada but also because of enormous costs. Much of the future of observational astronomy is in space; Canadian astronomers know this and, because of the historical development of the science in Canada, are uniquely prepared to join this entirely new phase in the history of astronomy. Only time will tell whether government will recognize this opportunity.

The real expansion of Canadian astronomical education did not occur until the 1960s and 1970s. Ontario institutions had always had the largest faculties and most extensive facilities, and this pattern continued, but other parts of the country increased their proportion of the whole significantly. In the Atlantic provinces, where astronomy had languished since the 1880s, new, small research centres developed at St Mary's, Acadia, and Memorial universities. Only the first of these has established a graduate program and, under David Dupuy's direction, an observatory, the small Burke-Gaffney, with its 16-inch telescope. The problem for eastern universities has always been their relatively small sizes. Astronomy is almost never seen as an essential part of a science faculty, being added typically only when the other sciences are sufficiently large in terms of staff and student demand or, rarely, through the extraordinary exertions of individuals.

In Quebec, where the universities were larger, the lack of development in astronomy was due to a different problem: the general weakness of the sciences. It was only in the 1920s that science was developed seriously at Laval and the new Université de Montréal. Astronomy, which emerged from the physics departments of the two schools, had to wait until the 1960s. Laval maintained a small observatory with a 41-cm reflector at St-Elzéar de Beauce, but Montreal, with specialization in theory and laboratory work, had none. By the 1970s, when demand and sufficient staff were present at both schools, the need for a research-calibre installation became pressing. Both groups collaborated on the design of the facility and, led by René Racine of Montréal, convinced the NRC to provide funds for the Observatoire astronomique du Québec.[46] The new observatory, the first major centre built in Canada since 1935, is located on Mont-Mégantic in the Eastern Townships and is equipped with a 1.6-m cassegrain reflector (f/15 cassegrain, f/8 Ritchey-Crétien). These developments ensured that graduate education in astronomy would be available for francophone students.

Queen's University had seen little change after the Second World War, but the pace quickened in 1955 with the appointment of George Harrower to the staff. He had worked with the Defence Research Board and introduced into the curriculum the study of radio astronomy. A small field station was built near Kingston in 1957 and eventually comprised aerial systems for the study of scintillating sources and solar emissions.[47] Although a 15-inch telescope was purchased soon afterwards, the astronomy group has concentrated on radio astronomy and theoretical astrophysics. Radio astronomy observations were eventually undertaken at the ARO.

Apart from Toronto, the other important astronomical centre in Ontario has been the University of Western Ontario. Under William Wehlau's supervision, a department of astronomy was formed during the 1960s and the 1970s, while laboratory astrophysics, directed by R.W. Nicholls, was prosecuted by the Molecular Excitation Group. The Hume Cronyn Observatory on campus was too limited for research, and negotiations with NRC led to funds for a new facility at Elginfield, north of London, equipped with a 48-inch instrument similar to the telescope at Victoria.[48] At both Queen's and Western, astronomy grew over a long period from earlier attempts, but during the 1960s, with the rapid growth of university education in Ontario, some of the newer universities included astronomy from the beginning. York University's Centre for Research in Experimental Space Science, established by Nicholls from Western, K. Innanen, and others, has become the largest single group besides the University of Toronto, although most of its work is in laboratory astrophysics and upper atmosphere studies. The University of Waterloo and Laurentian University in Sudbury both created small astronomy groups. Many of the staff members in Ontario astronomy departments or groups have come from Toronto, which has continued to be the major focus of astronomical education.

Toronto's department reached full maturity in the 1950s.[49] Under Frank Hogg's headship, a PH D program was finally instituted in 1947, and a trickle of graduates, beginning with William Hossack, Ian Halliday, and J.B. Oke, began to fill posts at Toronto and other Canadian universities and government agencies. There was no great swell of graduate students, however, until the 1960s, during the tremendous upsurge of interest in astronomy in Canada, as elsewhere. The introduction of the doctorate also signalled a widened scope in instruction. R.E. Williamson, a theoretician and radio astronomer, arrived in 1947, and Donald MacRae, a former Toronto student, returned in 1953; radio astronomy soon became a permanent feature of the programme, and several experimental antennae were built at the DDO. Eventually this work was transferred to the radio-quiet Algonquin site.

Although spectroscopy continued as a major focus of the department, photometry grew apace, requiring a new 24-inch reflector, with funds provided by philanthropist Walter Helm. One great area of expansion of optical astronomy during the 1960s was the creation of a number of southern-hemisphere observatories. Canada had no southern 'window,' until Toronto established itself at the Las Campañas Observatory in Chile with an Ealing-built 24-inch reflector. In fact, before the CFH telescope began operation, the southern telescope was the means of observation for more articles in prestige journals than any other. It was at Las Campañas that the first naked-eye supernova discovered since the seventeenth century was seen in early 1987, by Toronto's Ian Shelton.

Although the DDO, now fifty years old, is surrounded by urban sprawl, with most of its observational programs doomed by increasing light pollution, it remains the single most important centre in Canada for training astronomers. The Department of Astronomy is as large and diversified as any in North America, with specialists, on three campuses, in galaxies, cosmology, spectroscopy, photometry, radio astronomy, variable stars, interstellar matter, and several theoretical areas. The members of the department have provided an annual forum – the June Institute – to bring together outstanding specialists from home and abroad to present research findings on the latest fronts. With its graduates scattered throughout the world, Toronto, should continue, for the foreseeable future, as Canada's premier institution for both undergraduate and graduate training in astronomy. Space does not allow for a detailed account of the oldest astronomy department; a history of the department and the observatory would fill a gap in our knowledge of higher education in Canada.

In western Canada, only the older schools, Alberta and UBC, developed astronomy in depth. Radio astronomy figures largely at both institutions, the former operating a facility at Seven Mile Flat for long-wave radiation, the latter operating millimetric equipment on campus. Neither Saskatchewan nor Manitoba progressed much beyond the level of the 1930s. The newer universities in the west, as in Ontario, were more likely centres, and astronomy took root at Simon Fraser, Victoria, Calgary, and Brandon.

The western schools, too, have developed a wide variety of research lines, including infra-red, X-ray, and cosmic ray astronomy, along with more traditional spectroscopic and photometric work. Thanks to lower costs for modest-sized telescopes and the wide availability of electronic auxiliary equipment, most universities have been able to obtain a complement of instrumentation, and small observatories serve Brandon, Alberta, Calgary, Lethbridge, British Columbia, and Victoria.

A continuing problem in Canadian astronomy has been the oversupply of manpower as opposed to professional opportunities. This situation was not critical until after the Second World War, since the majority of graduates were from Toronto, and the number was never great. As numbers rose, however, openings did not keep pace; in the late 1960s, there was an influx of graduate astronomers to fill the rapidly opening positions in universities and government. None the less, many noted graduates of Canadian schools went elsewhere – chiefly to the United States, where the many more universities, private industry, and NASA offered far more possibilities, either immediately on graduation or after a few years in Canadian posts. In the nineteenth century, this 'brain drain' was very small, the best-known migrants being Simon Newcomb, S.A. Mitchell, and James Watson, but more recently this group has included H.H. Plaskett, A.B. Underhill, J.B. Oke, R.F. Christy, A.G.W. Cameron, W. Buscombe, D. Morton, D.E. Hogg, B.E. Turner, P.J.E. Peebles, and R.H. Hardie, who have made a significant impact in the United States and elsewhere. The Canadian government has been and is still the largest single employer of astronomers, but the university astronomers as a group are the largest bloc. Unlike the United States, Canada has no significant market in private industry, nor has it had a large-scale space agency.

The university situation at the end of the 1960s, when the first fruits of the 'Sputnik Effect' were manifest, was detailed in a 1969 study by R.C. Roeder and P.P. Kronberg for commission 46 of the IAU.[50] Of the forty-three universities surveyed, twelve offered no astronomy at all, eleven had one undergraduate course, and another four had more than one course but no graduate training. Only eight had both undergraduate and graduate programs: Victoria, British Columbia, Western Ontario, Waterloo, Toronto, York, Queen's, and Montréal. This group was represented by forty-three faculty members, twenty-seven PH D students, and forty-six master's-level students. In other words, a nation of more than twenty million people had only seventy-three graduate students in astronomy; some were non-Canadian, but, presumably, some Canadian students were studying elsewhere. The problem had a double aspect: the number of posts and programs had increased significantly, though not, perhaps, as much as some hoped; and the graduate student numbers, though small, already were too large to be absorbed into the educational system within the foreseeable future.

The situation improved somewhat during the ensuing decade. A 1978 survey of undergraduate education in astronomy showed some forty-one universities with at least an introductory course, along with twenty-four offering advanced courses.[51] Ten maintained undergraduate major programs, but little had changed at the graduate level.[52] Master's programs were offered by St Mary's, Memorial, Guelph, and the Université du Québec à Trois-Rivières, and both master's and PH D were provided by the same universities as in 1969, with the addition of Calgary, Alberta, and Laval. By 1977, Canadian universities employed some 100 astronomers or physicists working on astronomical problems. The graduate programs accounted for some 150 students. Since there has been no significant expansion of astronomical education in the 1970s or 1980s, the prospects for graduates continue to be difficult, but, except briefly in the 1960s, this has always been true in astronomy everywhere.

While most university-based astronomers work with optical and radio observations, a small number of theoreticians are scattered across the country. There being no Canadian 'think-tanks' devoted to astronomy, theoreticians have had to work in isolation. In the 1970s, CASCA encouraged short-term summer 'institutes,' some for theoretical work, at a cabin outside Toronto, and Queen's University, a centre for theoretical work, sponsored a series of national meetings. R.N. Henricksen of Queen's and others pressed for a permanent institution. Discussions brought together several universities, the NRC and NSERC, leading to establishment of the Canadian Institute for Theoretical Astrophysics in 1982. Funded by NSERC, the institute's home is in the same building as Toronto's Department of Astronomy.

The years immediately after 1945 saw no dramatic upsurge of popular interest in astronomy. That came, as everyone is aware, after the 1957 launching of Sputnik 1. Unfortunately, Canadian astronomers have been unwilling to write textbooks and popular works. Chant's little book, *Our Wonderful Universe*, first published in 1928, was successful and went through a number of printings and translations. After the war it was too dated for the changing market, and American and British books monopolized the bookshops and libraries. On the more technical level, one thinks immediately only of D.W.R. McKinley's *Meteor Science and Engineering* (1961) and Alan Batten's *Binary and Multiple Systems of Stars* (1973).

During the last half-century, among the daily press, the *Toronto Star* took astronomy most seriously. Astronomy columns by Peter Millman,

Frank Hogg, and Terence Dickinson framed the remarkable thirty-year (1951–81) tenure of Canada's pre-eminent astronomy popularizer, Helen Sawyer Hogg. She also provided one of the small but growing number of popular books, *The Stars Belong to Everyone*. Dickinson's *Nightwatch* added to an already distinguished career in science journalism, with articles in *Astronomy* and the Canadian natural science and exploration magazine, *Equinox*. DDO director Donald Fernie's historical sketches, *The Whisper and the Vision*, and books by well-known amateur astronomers Jack Newton and David Levy show wide diversity.

Overall, however, the print media in Canada have had an abysmal record in science reporting. Mass-circulation magazines have ignored science; newspapers, with notable exceptions such as Toronto's *Globe and Mail* and Montreal's *La Presse*, report only major scientific events. This has been true throughout the century. No purely science magazines of the calibre of *Scientific American* or *New Scientist* ever emerged. Those wishing news of Canadian astronomy might rely on the *Journal* of the RASC, although its record for reportage has been inconsistent since the war; the same has been true in recent years, for the (now-defunct) *Science Forum*, *Québec-Science*, and the NRC house journal, *Science Dimension*. University magazines occasionally report astronomical developments of their particular schools. Unfortunately, none of these organs has ever reached more than a tiny fraction of the Canadian public. It is often argued that because of the small potential readership in Canada, the print media could not afford to publish specifically science-oriented journals or carry much scientific news. It might be closer to the mark to ascribe this blinkered attitude to the provincialism of the media in Canada. This has been a classic chicken-and-egg problem for astronomy; few astronomers have bothered to write for the popular press, and there have been few opportunities for astronomers to write, even if they were so motivated.

Radio in Canada took some interest in astronomy during the 1930s, but much less since then. The rapid growth of private radio stations and networks, with emphasis on musical programming, meant that even marginally intellectual programming was driven back to state-controlled radio, the CBC / Radio-Canada. The state network, like its British counterpart, has a reasonable record for bringing science to the listening public, especially with the CBC *Ideas* and *Quirks and Quarks* series. Unfortunately, its listening audience is so small, and composed of the highly educated portion of the population, that astronomy programs rarely reach more than a few thousand.

Television, unlike radio, has tremendous possibilities for the popularization of astronomy through visual effects. But here, too, the finest astronomical programs have been imports from the BBC and publicly supported US television stations. A few Canadian astronomers have appeared on Canadian television, although often in local broadcasts as opposed to national programs. For example, the Toronto Centre of the RASC began production, in 1980, of 'Astronomy Toronto,' on a local cable network. In the 1950s, Canadian astronomy offered little that would excite public interest; but from the 1960s to the present, when Canadian astronomers participated in a great variety of exciting research, the record is little better, despite astronomy segments on CBC's *The Nature of Things*, and TV Ontario's *Fast Forward*, during the 1980s.

The medium of film has gone largely unexploited, apart from the National Film Board's ground-breaking *Universe* in the 1950s and NRC's film *To the Edge of the Universe*, on the Very Long Baseline Interferometry experiment of 1967. The former is perhaps one of the finest science films ever produced and features special effects still seen in science-fiction films.

Organizationally, Canada has not kept pace with the United States in finding means for popularizing astronomy. By 1939, the RASC counted nearly 1,000 members; figures took a predictable drop during the war but surged to 2,016 by 1948. Yet, by 1985, after the Canadian population nearly doubled, the society had reached only 3,500. In Quebec, small local clubs were linked with RASC centres in 1975 as L'Association des groupes d'astronomes amateurs (AGAA), later to publish *Le Québec astronomique*. During the 1970s, public interest in the United States increased tremendously and was reflected in the growth of astronomy clubs throughout that country. There are a few clubs in Canada besides the twenty RASC centres, but nowhere near the widespread interest seen south of the border. Amateurs who might attract public interest, such as comet discoverers Rolf Meier and David Levy, are rarities.

There are several explanations. First, much of Canada's post-war population growth was fuelled by immigration, and it might be argued that immigrant groups are not so likely to take up astronomy as an avocation. Second, and perhaps decisively, amateur astronomy may not be appealing to many but a few stalwarts in a climate such as Canada's. (Note, for example, that in the United States, two great areas of amateurs are the southwest and west, both blessed with reasonable climates and access to clear skies.) Couple that with no indigenous industry for astronomical instruments for the amateur and the high price of imports, one is unlikely to see telescopes on every street. The RASC has always relied on its professional and dedicated amateur elements for survival; the 'armchair' astronomer would never be a significant force in Canadian societies.

Nevertheless, the RASC has endeavoured to bring astronomy forward with public presentations and star nights. Since the war years, the number of centres has increased, with new organizations in Calgary, Halifax, Kingston, Niagara Falls, St. John's, Saskatoon, Sarnia, and Windsor. Both the Kingston and St John's organizations began as small clubs connected with the local universities. Saskatoon's centre had a false start (1947–52) before re-establishment. The francophone members in Montreal formed the separate Centre d'astronomie in 1947, but it has remained much smaller than its anglophone counterpart. One move to foster local awareness in other parts of Canada was undertaken in 1958 by making the annual meeting peripatetic. The general assembly has become a significant amateur gathering.

The *Journal* of the society had become less and less relevant to amateur interests after the war. Since it was an outlet for professional papers, few amateur items appeared. As Canadian astronomy grew in breadth, however, so did the *Journal*. The first issue of 1948 was something of a harbinger, with an article on radio astronomy by Ralph Williamson and another on space travel by J.W. Campbell. Since then it has reflected the growth of modern astronomy. During the 1960s brief articles, RASC papers, given at general assemblies, began to be included, some of purely amateur interest. The 'Variable Star Notes' from the American Association of Variable Star Observers, a long-time feature, is also used by amateur specialists. At the 1969 general assembly, it was decided that a special insert, the 'National Newsletter,' should be added. This feature was initiated in the January 1970 number. The newsletter carries centre news and items of amateur interest. Since 1978 this feature has been published separately. A number of centres have also produced their own newsletters. Although amateur interest is the core of the modern RASC, the professional element has not disappeared, even with the formation of CASCA.

The society, especially through its centres, took a leading role in the creation of the most important public monuments to astronomy, planetariums. The planetarium came late to Canada. The Zeiss projector, created in Germany in the 1920s, was too expensive for Canadian cities, and the first North American planetariums were located in large American cities such as New York and Chicago. These were the envy of Canadian astronomers who visited them during professional meetings. The situation changed after the war with the invention by Armand Spitz of a smaller, less sophisticated, and cheaper instrument. Members of the Hamilton Centre of the RASC, visiting the Buffalo planetarium, were impressed and decided to purchase a Spitz projector for McMaster University. They obtained a $1,000 instrument and built a small dome from a used parachute in November 1949. A permanent dome was added in 1954. A year earlier the RCAF installed a Spitz machine at Base Summerside, PEI.

Until the 1960s, the only other planetarium opened was in Halifax. This was a project of the Nova Scotia Astronomical Society, forerunner of the Halifax Centre, RASC, founded in 1951. A Spitz AI instrument was purchased, and a small planetarium was erected in the Nova Scotia Museum of Science in 1956. With increasing public awareness of astronomy, the idea of having a planetarium struck many civic authorities during the 1960s.

The first of the modern, municipal plantariums was the Queen Elizabeth in Edmonton, built to commemorate a royal visit. It opened in 1960. This was followed in the mid-1960s with major planetariums in Montreal, Winnipeg, Toronto, Calgary, and Vancouver. The planetariums in Calgary and at the Manitoba Museum of Man and Nature were centennial projects, while those in Toronto, Montreal, and Vancouver were provided through private philanthropy. With more money available, most were able to purchase Zeiss instruments. Universities and colleges have been much slower, only the University of Manitoba and Seneca College in Toronto having full-scale facilities. With nearly a dozen installations in the country, a Planetarium Association of Canada was founded in 1965 for the exchange of ideas.

Canada is not well endowed with science museums, and those large enough to cultivate astronomy do so on a small scale. The National Museum of Science and Technology in Ottawa has a small planetarium and in 1974 acquired the DO's historic 15-inch refractor from EMR. With these instruments, and a small exhibit on astronomy, the museum's astronomy curator, Mary Grey, has pressed on with a vigorous public program. Science North, in Sudbury, and the Ontario Science Centre, in Toronto, have developed modest astronomical exhibitions. Western Canada was represented by the new Space Sciences Centre in Edmonton. The weakest areas of the country for popular astronomy continue to be the Atlantic region and French Canada. It is evident that the cultivation of astronomy by the public requires relatively large populations, but the Atlantic provinces, with small towns and modest cities, have never been able to sustain major museums and planetariums. French Canada has only recently begun to see science as a respectable vocation, and its modernized school curricula is only some twenty-five years old.

Post-1945 Canadian astronomy stands out in stark relief when compared with what went before. During the 1930s, Canadians had gained some stature in the international astronomical community but were still small in number, persuing as yet few lines of research. By the late 1970s, the overall picture began to resemble that of any mature scientific nation, such as the United States or the United Kingdom, except that the scale was smaller. Wartime work was responsible for new technological capabilities and new fields, radio astronomy and astronautics being obvious examples. Because Canada was tied into the war effort as a relatively senior Allied partner, its scientists were able to participate in major research programs such as radar and nuclear physics in a way that other nations of similar size could not. Its proximity and similarity to the United States were a great advantage. By 1945, Canada, like other major nations with significant wartime scientific organizations, had available a sizable complement of highly trained and skilled scientific workers, some of whom were astronomers or physicists who could move easily into astronomically related problem areas. The NRC, which eventually amalgamated all of government astronomy under its aegis, increased an order of magnitude in size during the war and retained most of its diverse programs and staff after the end of hostilities. Returning veterans swelled the universities. This wartime and immediate post-war boom increased the scale of science in Canada, and this was not dismantled later.

The mid-1950s brought a different phenomenon, which accelerated the interest in the physical sciences in general, and in astronomy in particular. This is what I have called the 'Sputnik effect,' the psychological stimulus to educational reform and government spending on science at an unprecedented rate. The so-called space race was, in fact, the equivalent of a war effort, and Canada, with all the close ties forged before and during the Second World War, was not immune to this primarily American phenomenon. Canadian astronomers naturally strengthened their ties with Americans, because of family resemblance – both being North American societies – and proximity. Also, the American astronomical community was and is the world's largest and richest in terms of instrumentation and programs. No astronomer can ignore American astronomy, and many Canadians were educated in the United States and maintained close contact with American colleagues and published in the numerous American journals. The broadening of research interest in Canada cannot but be affected by spill-over from the United States.

One feature of Canadian astronomy from earlier times did remain, however – leadership. The individual counts more in a small community than in a larger, and the outstanding astronomer in Canada is more visible within his or her limited context than remarkable practitioners would be in a large community such as the American. Chant and Plaskett had both effectively retired as leaders by the mid-1930s, but their successors, men such as R.M. Petrie, C.S. Beals, and P.M. Millman, continued to have a strong effect upon the field. The leadership factor probably receded in importance in the late 1960s and the 1970s: the size of the community, the number of institutions, and the increasing bureaucratization of science by government have blunted the role of the individual. Indeed, this shift from the individual leader to the collective may be a sign of maturity in a scientific community.

Finally, an important feature of post-war Canadian astronomy has been the enhanced international profile of Canadians. There is no doubt that astronomers, no matter how brilliant or dedicated, devoted to practical or routine work, rarely gain notice beyond a small group of specialists in their area. The far wider scope of Canadian astronomy since the war has allowed more Canadians to participate in a variety of programs with international impact. The reputation of Canadians doubtless was a factor in the selection of Montreal as the site of the 1979 general assembly of the IAU. The increasing notice of Canadian work by the international community must reinforce the growth of sub-disciplines in Canadian astronomy.[53] This growth has slowed and stabilized since the ebullient 1960s, but Canadian astronomy should continue to be fully tied into international research.

Conclusion

Few historical changes are sudden: they can be seen in embryonic form long before they appear as obvious change. So it was with the maturing of twentieth-century Canadian astronomy. Already, by mid-nineteenth century, astronomy in Canada had given way to Canadian astronomy – a definable group of men rooted in Canada and employing their science to fulfil indigenous needs. In the first few decades of the twentieth century the science matured in terms of manpower, educational base, choices of research, funding, national and international recognition, and, most of all, the subtle psychological shift in astronomers' minds that allowed them to see themselves as professionals and the equals of foreign scientists. Such a process does not occur in a vacuum, and not all scientific groups emerge as professional. The social and economic context must be equally advanced before a science can mature and survive. Therefore, Canada's socio-economic trajectory matched the growth of astronomy.

Most of modern astronomy is a luxury to society. The practical side – calendar and time reckoning, navigation, and geography – requires few specialists and is relatively inexpensive. Astrophysics, the most significant twentieth-century manifestation of astronomy, is of little value to society at large and even less value to the economy, despite arguments that pure science leads to technology. Indeed, it can be a substantial drain on taxpayers, given the small numbers who benefit from the financial support to astronomy. Precisely because of the small social and economic benefit of modern astronomy, as opposed to that of chemistry, biology, or physics, support for it acts as an important indicator of the maturity of the society itself. Just as early human societies that supported separate groups of clergy, bureaucrats, and artisans were more culturally, socially, and economically advanced than those that did not, so we can judge contemporary societies by the extent to which they support apparently unproductive activities such as the fine and performing arts and the pure sciences. Thus, by the early years of this century, Canada's cultural maturity was sufficiently advanced to support astrophysics.

The emergence of science is part of the diversification of societies, and the history of astronomy is a small facet of that process. The rate of diversification depends on many factors, including the influence of neighbouring societies. This factor can work on both the large scale and on the small; the history of American astronomy has strong similarities to that of Canadian because both societies diversified along much the same lines and felt mutual influences. Canadian astronomy developed more slowly because Canadian diversification was slower. Both countries were peopled by similar groups of immigrants, but the population bases were always substantially different. The Canadian land, as vast and rich in resources as the American, is far more intractable for settlement, its climate more rigorous, its agriculture harder to develop, and, most important, its industrial base much slower to emerge. Without a large industrial sector, a country cannot produce a large enough middle class – the class of science – or aggregate enough of its members into large urban centres – the location of much innovation – to support large-scale scientific efforts. Canada began to overcome these barriers to development only toward the end of the last century. Not surprisingly, Canadian society could not diversify to the same extent as American and could not support so rich a variety of scientific activity, save that with distinct practical or economic value. Astronomy and, for that matter, physics fell into this category in nineteenth-century Canada.

Canadian astronomy had developed in small steps during the nineteenth century. While the number of professional and serious non-professional astronomers increased very slowly from 1840 to 1905, growth was relatively steady as in the other physical sciences. Small observatories had been erected across Canada, but by 1905, their very smallness in terms of financing, personnel, instrumentation, and programs had doomed them to extinction. Brydone Jack's and Joseph Everett's observatories were unused or in ruins; the Quebec Observatory no longer contributed to research; the Queen's Observatory was silent; and the Toronto Observatory was busy with meteorology. The older generation of astronomers was passing: Ashe and Williamson were dead in 1895, Dupuis and McLeod died in 1917. The new core of Canadian astronomers were government men: King, Klotz, Deville, Stewart, and Plaskett.

The older, pre-1905 system began largely with astronomy but ended in meteorology. The roots grew in the soil of practical need. A new country needed men to find latitudes and longitudes and to provide the time; the solar work of Ashe and McLeod was a small luxury. The observations of comets, of the transit of Venus, or of solar eclipses were chances for eminently practical men to participate in the greater excitement of nineteenth-century astronomy. However important their daily work, these men sensed that events of this nature allowed them, for a brief time, to feel more a part of the international fraternity of astronomers. Men like Ashe, McLeod, and Williamson were proud of their efforts.

During the nineteenth century there were no important discoveries and no great astronomers, but the few astronomers there were did provide the essential astronomical contributions to their developing nation. Further, the astronomy of Canada became Canadian – its growth pattern was different from that of other nations – and it was not an adjunct of British science, nor merely a pale imitation of American science, nor, for the most part, even a link in imperial science, despite some of the rhetoric that occasionally saw print.[1]

The Canadian government's involvement in astronomy was far greater relatively than the British or American, not only in astronomy but in most of science as well. Government nurtured the first system of Canadian astronomy and established the second. Given the resources and outlook of Canadians, it could have happened no other way. No other institution had the wherewithal to foster astronomy.

Canada began to diversify more rapidly by the turn of the century. Industry was spreading, the railway network in place, the west being rapidly colonized, the population increasing, the economy improving, and two major cities, Toronto and Montreal, emerging as centres for the sciences. As mining, agriculture, and industry expanded, the need for science increased. Both university and government sectors responded: the period from about 1885 to 1920 saw unprecedented institution-building in science and technology. Astronomy was part of this movement, with government creating the two national observatories, and the University of Toronto establishing its Department of Astronomy.

Diversification operates in societies through the emergence of new social groups, industries, and institutions; in science, through the emergence of new specialties, institutions, facilities, journals, and societies. Canadian astronomy about 1900 had diversified to the point that one major new specialty, stellar astrophysics, secured a footing. The period up to the Second World War was largely one of consolidation. None the less, a few new lines of research opened up, a major educational observatory was founded, a national society and journal were created, and manpower slowly increased. But two wars and a serious depression dislocated Canada's growth pattern of the turn of the century; astronomy could only follow suit.

When the Second World War ended, however, diversification began anew. Hundreds of thousands of immigrants arrived annually, the economy boomed for some twenty years, and the 'Sputnik era' put tremendous pressure on the educational system. All these factors helped to establish a climate for further scientific growth and diversification. Canadian astronomy began to resemble astronomy in other highly developed nations in terms of breadth of activity. The chief difference between Canadian and American astronomy, for example, became largely one of scale.

The professionalization of Canadian astronomy took place long before the latest wave of diversification. The process was probably under way when astronomy became part of the civil service in the late nineteenth century and became firmly established once the University of Toronto began graduating people who could move directly into an astronomical career. The process was certainly complete by 1930. One measure of that professionalization is the extent to which Canadian scientific work was acknowledged and actually employed by ranking scientists of more developed nations. One cannot feel an equal to colleagues elsewhere if one's work is not considered equal; the evidence in appendix B suggests that the recognition of Canadian astronomical research differed in no substantial way from that of other developed astronomical communities by the 1920s.

The development from the colonial context to mature, national status, and the consequent evolution of Canadian astronomy from a series of isolated, unco-ordinated, and little-recognized undertakings to a mature, professional science, have obvious resemblances to the evolution of other countries. Although the process of diversification and maturation is much the same everywhere, the specific factors remain different. The history of Canada itself is unique. The internal social structure of the Canadian astronomical community, the role of governments, the form of education, the choices of research, the areas of competence, and the role of individual and collective all are as they are because of the specific way in which the country has evolved. The story of the development of Canadian astronomy covers a brief period of history and deals with very few men and women, but it is a proud and revealing one.

Observatories before 1900

No observatory of international importance existed in Canada until the establishment of the Dominion Observatory in 1905. Several of the nineteenth-century observatories had passed into total oblivion by the time that Otto Klotz penned a survey of earlier institutions in 1918–19; he listed only six observatories antedating the Ottawa establishment. We now know that perhaps as many as sixteen (plus or minus two or three) can be enumerated before 1900.

There is still some confusion and controversy over which observatory could be called the first in what is now Canada. For several decades, the prime candidate was William Brydone Jack's observatory at the University of New Brunswick, championed by Prof Ed Kennedy, who, more than anyone, brought Jack and his work to Canadians' attention. More recent scholarship has suggested that eight observatories were earlier than the Fredericton site. Let us review the candidates.

Collège de Québec (1750s?): We have no direct evidence that an observatory was erected for the college, although we do know that the professor of hydrography, Joseph de Bonnécamps, sj, who served at Quebec 1750–9, had several astronomical instruments, including a quadrant, clock, and small refracting telescope, and that he intended to build a permanent enclosure for them. A water-colour of the college, painted after the 1759 bombardment of the city, does not indicate a roof-top installation or an outbuilding for astronomical purposes.

Séminaire de Québec (?): Facing the college (now the site of the Hôtel de ville) was, and is, the Séminaire, which took over the teaching functions after the college was permanently closed by the British authorities. There is one mention (1770) in the seminary accounts of the construction of a 'chambre d'observasion,' perhaps a small observatory. Research under way concerning the early scientific instruments at the Séminaire may provide further elucidation. There was, however, no mention of an observatory in the early nineteenth century.

Louisbourg (1750–1?): R.C. Brooks has argued that the marquis de Chabert de Cogolin, while in the Louisbourg area 1750–1, built an observatory within the fortress. Chabert brought with him a quadrant, a clock, an octant, six refractors, and one reflector. The only contemporary woodcut, from Chabert's book, shows observers with the quadrant out-of-doors. The building, which seems to have existed for a short time, was probably akin to the temporary telescope shelters built by Department of the Interior observers in the 1880s and 1890s. There is no evidence that an observatory survived Chabert's return home.

Castle Frederick Observatory (1765): Roy Bishop has documented the establishment of an observatory in Falmouth, N.S., by the noted surveyor and hydrographer Joseph DesBarres, in 1765. Equipped with quadrant, Hadley's quadrant, reflector, and refractors, the observatory appears in a contemporary picture. Bishop claims that this observatory was the first in North America.

Toronto Magnetic Observatory (1840): This observatory, founded in 1840 by the imperial government, was intended not for astronomy but for geophysics. We do know that the observatory eventually possessed a small transit instrument and clocks, but the evidence does not suggest any astronomical observations before the 1850s. By the 1880s, with the addition of the Cooke refractor, some limited work was performed and the instruments were placed at the disposal of the University of Toronto.

Île Jésus Observatory (1846?): The small establishment of Dr Charles Smallwood at St-Martin, Que., was described in detail in 1858; while its purpose was essentially meteorological, there was a slit in the roof, a Fraunhofer equatorial, and a small transit instrument. Smallwood did make some astronomical observations, and there are hints that the observatory was built as early as 1846.

James Toldervy Observatory (1849?): We know very little about Dr Toldervy except Brydone Jack's mention of their collaboration in the 1850s. Jack does say that Toldervy had a private observatory in his garden, near the river in Fredericton, and had at least a clock and a 30-inch Simms transit. The evidence points to at least as early as 1849 for establishment.

Quebec Observatory (1850): The traditional date for its founding is 1855, but, in fact, the building was finished and occupied by E.D. Ashe in the autumn of 1850. Although the time-ball machinery was not in place until 1852, the clocks and other instruments were in operation in 1851.

Therefore, we have eight candidates for an observatory ante-dating the erection of Brydone Jack's building and telescope in 1851. Of these eight, the earliest two, in Quebec are conjectural; Louisbourg was perhaps an observing station rather than a permanent observatory; and Toronto was probably not astronomical until much later. This leaves Castle Frederick, Île Jésus, and Toldervy's all private, along with the state-operated Quebec Observatory, as those antedating Jack's. As we lack sufficient information about the establishment of Smallwood's and Toldervy's observatories, only Castle Frederick and Quebec stand with full documentation. I think it likely, however, that all four were in operation before 1851.

Between 1851 and 1900, a further eight permanent observatories are definitely known to have been erected. There may have been further amateur installations, possibly adding another two or three to the total. Information concerning these later institutions is in the text, and here they need only be listed with the dates of establishment:

Kingston Observatory, 1856
King's College Observatory (Windsor, NS), 1862
McGill College Observatory, 1863
Charles Blackman Observatory, Montreal, 1873?
Woodstock College Observatory, 1879
Victoria College Observatory, before 1882
Cliff Street Observatory, Ottawa, 1890

All the pre-1900 observatories have long since ceased to exist. Castle Frederick was probably gone by 1783, and there are no traces of observatories in Quebec (if they existed at all) or Louisbourg. The Île Jésus instruments were moved to the McGill campus after 1862; Blackman's instruments joined them in 1879. The original McGill Observatory was razed in 1963 but had long been unused for astronomy. The Quebec Observatory, either in the Citadel or at the Bonner Farm site, was gone sometime between the two world wars. The original Toronto Observatory site was abandoned in the 1890s, although the second building survives on the University of Toronto campus, as does the third, nearby on Bloor Street. There remains no trace of Toldervy's observatory, but Jack's university observatory, with some of the original instrumentation, survives as an historic site. The original Kingston Observatory, and its campus successors, were gone by the early twentieth century, although the Clark refractor survives. The three college observatories, King's, Victoria, and Woodstock, were all effectively defunct by 1900, although a little work may have been done in the last as late as about 1910. They have all disappeared. The Cliff Street Observatory was rebuilt in 1912 and was used until 1939; the Supreme Court building now stands on the site. The oldest, still active observatory in Canada is the Dominion Astrophysical Observatory, which commenced operations in 1918.

Education 1869–1950

The number of Canadian astronomers remained small until after the Second World War, as did the number of institutions that trained them. Below are tabulated the names of people graduated from Canadian institutions with first or second degrees between 1869, the year that W.F. King took his BA at Toronto, and 1950, just before the first PH D students began issuing forth from the same university. The list of names is reasonably complete and includes degrees granted in cognate fields such as physics and mathematics. Until well into the present century, most first degrees were in arts; the B SC or equivalent was a relatively late innovation. In 1900, seventeen universities granted degrees in Canada; the number rose to twenty by 1920. Fewer than half accounted for graduates who worked in some area of astronomy, and the dominance of Toronto is evident.

Sources of First and Second Degrees

Acadia University: Beals
University of Alberta: McDonald
University of British Columbia: Covington, McKellar, R. Petrie, W. Petrie, Richardson, Underhill, Volkoff
University of Manitoba: Cameron, Thomson
McGill University: Burland, Douglas, Heard, Kennedy, McLeod
Queen's University: Dupuis, Kennedy, Kingston, Warren
Royal Military College: Bigger
University of Western Ontario: Heard
University of Toronto: Arbogast, Balmer, Beals, Bunker, Cannon, Chant, A. De Lury, R. De Lury, Halliday, Harper, Henderson, E. Hodgson, J. Hodgson, Hogg, Hossack, Jaques, Johns, King, Locke,

McClenahan, F. McDiarmid, R. McDiarmid, McLean, McLennan, MacRae, Millman, Motherwell, Northcott, Oke, Parker, Pearce, H. Plaskett, J. Plaskett, Sheppard, Shrum, Smith, L. Stewart, R. Stewart, Tidy, Wright, Young

Those names repeated represent people taking a second degree at a different university. This list does not include Canadian astronomers who took degrees elsewhere (e.g. Klotz, Redman, Henroteau) or those with Canadian degrees whose careers were elsewhere (e.g. Mitchell).

The number taking a PH D or equivalent is, of course, much smaller. The doctorate was not perceived to be a requirement for astronomy until about the 1920s. No Canadian university offered a doctorate in astronomy until the 1950s, and so Canadians had to go elsewhere; again the number of institutions is surprisingly small.

Sources of Doctorates

CANADA
Saskatchewan: Cameron*
Toronto: J. Hodgson,* McLennan,* Meen,* Shrum*

UNITED KINGDOM
London: Beals,* Heard*

UNITED STATES
California: Christie, McKellar, Pearce, Young
Chicago: Buchanan, Campbell, Kingston,* Underhill, Warren*
Harvard: Chant,* F. Hogg, H. Hogg, Millman
Michigan: R. Petrie, Wright
Princeton: R. McDiarmid

Those marked with an asterisk took doctorates in fields other than astronomy. Before 1950, few universities in the world offered a PH D in astronomy, but Canadians were concentrated in even fewer institutions because of various personal ties. Chant had close links with the Lick Observatory staff (University of California) and sent a number of his students there, although not all finished. He likewise directed students to Harvard, where he had taken his doctorate, once Shapley established a viable graduate program. Chicago was an important centre because of mathematics, attracting Queen's students Kingston and Warren, and because of the

Yerkes Observatory, where Underhill and Douglas – who did not take the degree – worked. Fowler's spectroscopic work at London attracted physics graduates Beals and Heard. At Toronto, Shrum was McLennan's student in physics, while Hodgson and Meen were earth science graduates.

Visibility 1920–1929

During the nineteenth century, Canadian astronomers worked at the margins of international astronomy; only in the twentieth century, when astrophysics became an important subject among Canadians, did their 'visibility' rise. One measurement of the visibility of scientists is the extent to which their work is cited by others. Historical records of citations have been organized in *The Physics Citation Index* covering the period 1920–9, which happily coincides with the first great burst of Canadian astrophysical research.

There are limitations and problems with the *Citation Index* that render any conclusions only approximate. First, only items cited during the period are listed; thus, in the tables below, total items refer not to the total publishing record of a person, but only to those items cited. Second, the index includes only citations in sixteen major international physics journals; for astronomy, the most important in the list are *Monthly Notices, Astrophysical Journal, Proceedings of the Royal Society – Section A,* and *Physical Review*. Many important astronomical journals are missed in this selection, but the journals used for citations were all highly prestigious. Third, the index lists publications and citations under authors' names but also sometimes lists only a surname. For example, the works of R.K. Young are listed under his name, but there are also many references under 'Young,' which include some by R.K. Young but also some by the dozen or so other Youngs listed. As it would be very difficult and time-consuming to work out which Young was which, I have ignored the general headings. The items listed include articles, observatory publications, monographs, as well as acknowledgments, and can include self-citations.

In Table 1, I have listed the sixteen Canadian astronomers who were cited during the ten-year period in the 'prestige' physics journals, the number of items cited, the number of citations, average citations per item,

TABLE 1
Citations of Canadian Astronomers, 1920–9

Name	Items	Citations	Citations / item	Prestige publications
Beals	6	18	3.00	6
Cannon	17	10	0.58	0
Chant	4	5	1.25	0
Christie	3	3	1.00	0
De Lury, R.	3	3	1.00	1
Douglas	15	20	1.33	8
Harper	66	43	0.65	1
Henderson	1	1	1.00	0
Henroteau	50	42	0.84	2
McDiarmid, R.	2	2	1.00	0
Pearce	14	14	1.00	0
Petrie	2	2	1.00	0
Plaskett, H.H.	24	29	1.21	6
Plaskett, J.S.	99	100	1.01	25
Stewart	4	3	0.75	1
Young	25	19	0.76	0

and, finally, publications in 'prestige' journals. Where the citation / item value is below 1.0, several items will have been cited collectively in one citation.

The average citation / item for the entire group is 0.94. We cannot read much into the specific values, given the idiosyncracies of the *Citation Index*. Beals, for example, has a very high citation / item rating: his early work, in physics, was associated with the laboratory of Alfred Fowler at London, giving Beals easier access to prestigious British journals.

By way of comparison, we can select American astronomers active in the same period (see Table 2). Rather than try to match up Americans with the Canadians in Table 1 by age, experience, or ability, I have chosen them by the 'random' means of selecting those who appear on the same pages as the Canadians in the *Citation Index*. Only eight names appear, but these include some of the best-known American astronomers of the time, such as W.W. Campbell, Annie Cannon, and Forrest Moulton, along with younger, active researchers such as Cecelia Payne.

The average citations / item for this group is 1.01. There are also a number of oddities in the index: Campbell's work is not complete, but the entry for 'Campbell' is like that for 'Young,' and I have enumerated only those under 'Campbell, W.W.' Pettit's citations are mostly self-citations, giving him a higher value. I have assumed, however, that the oddities in the American list cancel out the oddities in the Canadian.

TABLE 2
Citations of Selected American Astronomers, 1920–9

Name	Items	Citations	Citations/ item	Prestige publications
Campbell	14	12	0.86	0
Cannon	70	61	0.87	0
Douglass, A.E.	7	6	0.86	0
Moulton	19	15	0.79	2
Payne	94	94	1.00	1
Pease, F.G.	66	76	1.15	5
Pettit, E.	70	72	1.03	27
Stewart, J.Q.	29	38	1.31	21

Strikingly there is no sensible difference in the 'visibility' of Canadian and American astronomers in prestigious journals during the period. Half the Americans in Table 2 have citation / item values of 1.00 or above; more than half the Canadians in Table 1 do so. The Americans were no more visible than the Canadians, especially considering that no major journal was published in Canada.

The conclusion is obvious: by the 1920s, Canadian astronomy, although still very small in scale compared with American astronomy, had a similar claim to notice in the world's scientific literature. The group that had this visibility was the astrophysicists, a group that was non-existent in 1900.

Notes

Abbreviations

Ap. J. Astrophysical Journal
JRASC Journal of the Royal Astronomical Society of Canada
MNRAS Monthly Notices of the Royal Astronomical Society
MSRC Mémoires de la Société royale du Canada
PAC Public Archives of Canada, Ottawa
Phil. Trans. Philosophical Transactions, Royal Society of London
PRSC Proceedings, Royal Society of Canada
Pubs. DAO Publications, Dominion Astrophysical Observatory
Pubs. DO Publications, Dominion Observatory
THLSQ Transactions, Literary and Historical Society of Quebec
TRSC Transactions, Royal Society of Canada

Introduction

1 For a survey of differing views of Canadian history, see Carl Berger, *The Writing of Canadian History: Aspects of English-Canadian Historical Writing since 1900*, 2nd edition (Toronto 1986). The dynamic tension view is argued by Herschel Hardin, *A Nation Unaware: The Canadian Economic Culture* (Vancouver 1974).

2 I have outlined these, along with critiques of other views of colonial science, in 'Differential National Development and Colonial Science: The Cases of Ireland and Quebec,' in N. Reingold and M. Rothenburg, eds., *Scientific Colonialism: A Cross-Cultural Comparison, 1800–1930* (Washington, DC 1987).

3 The role of urbanization is discussed in Walter Hendrickson, 'Science and

Culture in the American Middle West,' *Isis*, 64 (Sept. 1973), 326–40; *cf* my article, 'The Rise and Decline of Science at Quebec, 1824–1844,' *Histoire sociale*, 9 (May 1977), 77–91.

4 For a short survey of the period, consult T.H. Levere and R.A. Jarrell, eds., *A Curious Fieldbook: Science and Society in Canadian History* (Toronto 1974).

Chapter One: Colonial Astronomy

1 Thomas James, *The Dangerous Voyage of Capt. Thomas James* (London 1633; reprinted 1740), 129

2 Ibid, 136–8

3 At those latitudes, the length of one degree of longitude is approximately 35 miles. For a detailed description of the methods and accuracy of the seventeenth-century arctic explorers, see Peter Broughton, *JRASC*, 75 (Aug. 1981), 175–208.

4 Sir John Ross, *Narrative of a Second Voyage in Search of a North-West Passage* (London 1835)

5 Henry Scadding, *The Astrolabes of Samuel Champlain and Geoffrey Chaucer* (Toronto 1880)

6 R.G. Thwaites, ed., *The Jesuit Relations and Allied Documents* (*JRAD*) (New York 1959), vol. 28, 143

7 See Roland Lamontagne, *Roland-Michel Barrin de la Galissonière 1693–1756* (Quebec 1970). Geology and botany were of particular interest to several officers and officials during the 1740s and 1750s, but the French government never undertook a survey of the colony's natural resources or underwrote systematic exploration.

8 A recent study by Brooks claims that Chabert's observatory at Louisbourg was the first in North America. The evidence for this seems equivocal. See R.C. Brooks, *JRASC* 73 (Dec. 1979), 333–48; see also appendix A.

9 Joseph-Bernard, marquis de Chabert de Cogolin, *Voyage fait par ordre du Roi en 1750 et 1751 dan l'Amérique septentrionale pour rectifier les cartes ...* (Paris 1753)

10 J.-D. Cassini, *A Voyage to Newfoundland and Salee, to Make Experiments with Mr. Le Roy's Timekeepers* (London 1778)

11 *JRAD* 5: 65–7. For further details on the accuracy of French observations, see Broughton, *JRASC*, 18ff.

12 *JRAD*, 5: 99

13 Ibid, 8: 63

14 Ibid, 12: 141–3

15 Ibid, 48: 37–9
16 Ibid, 48: 43–5
17 See John Eddy, *Science*, 192 (1976), 1189–1202.
18 *JRAD*, 48: 241
19 Ibid, 50: 75
20 Ibid, 54: 241–3
21 Ibid, 58: 183–5
22 C. Le Clercq, *New Relation of Gaspesia* (Toronto 1910), 360
23 *JRAD*, 60: 205–7
24 L.-P. Audet, *Histoire de l'enseignement au Québec* (Montreal 1971), 1, 31ff
25 R.S. Harris, *A History of Higher Education in Canada 1663–1960* (Toronto 1976), 14
26 Audet, *Histoire de l'enseignement*, 193
27 Ibid, 193–4. A survey of the course of hydrography may be found in Audet, *Cahiers des Dix*, 35 (1970), 13–37. A 250-page notebook on astronomy by Guillaume Brunet, dated 1677 has been found in the Archives du Séminaire de Québec by R.-N. de Tilly. This manuscript probably represents the course taught by Martin Boutet (Montreal, *Le Devoir*, 1 August 1973)
28 P.-G. Roy, *MSRC*, Ser. 3, 13 (1919), 47ff
29 Audet, *Histoire de l'enseignement*, 201
30 See Auguste Gosselin, *MSRC*, Ser. 2, sec. 1 (1895), 25–61; (1897), 93–118; (1898), 33–5
31 Ibid, (1895), 27
32 'mais pouvais-je et devais-je compter sur une montre d'une bonté mediocre et dont j'ai cent fois éprouvé l'irregularité avant et après mon retour?' ibid (1895), 42
33 Ibid, (1897), 98
34 Pehr Kalm, *Peter Kalm's Travels in North America*, ed. A. Benson (New York 1966), II, 413–14
35 Antonio Drolet, *Revue d'histoire de l'Amérique française*, 14 (1960) 487–544. See also Drolet, *Naturaliste canadien*, 32 (Avril–Mai 1955), 105–6
36 Cited in G. Frégault and M. Trudel, eds., *Histoire du Canada par les textes* (Montreal 1963), 80
37 The most comprehensive survey is that of Don W. Thomson, *Men and Meridians* (Ottawa 1966), I, passim.
38 S. Holland to J.G. Simcoe, 11 January 1792, cited in Henry Scadding, 'Surveyor-General Holland,' *Canadian Magazine* 5, 6 (1885), 521–4
39 James Cook, *Phil. Trans.*, 57 (1767), 215
40 Samuel Holland, ibid, 58 (1768), 45–53 (in two parts); ibid, 247–52; ibid, 64 (1774), 171–6

41 See Roy Bishop, *JRASC*, 71 (Dec. 1977), 425–42.
42 William Wales and Joseph Dymond, *Phil. Trans.* 59 (1769) 467–88. See also Helen Sawyer Hogg, *JRASC*, 41 (1947), 319–26; ibid, 42 (1948), 153–9, 189–93; ibid, 44 (1950), 123.
43 H.C. King, *The History of the Telescope* (London 1955), 82
44 Samuel Hearne, *A Journey from Prince of Wales's Fort in Hudson's Bay to the Northern Ocean* (London 1795)
45 Alexander Mackenzie, *Voyages from Montreal ... to the Frozen and Pacific Oceans in the Years 1789 and 1793* (London, 1801)
46 For a survey of early observations in the Northwest, see J.S. Plaskett *JRASC*, 77 (June 1983), 108–20
47 PAC, MG 12, DI2 T28, V. 14, 15: Great Britain, Treasury, Out Letters 1816–17
48 A manuscript of astronomical problems composed and solved by Bayfield survives in PAC, MG 24, F28, V. 2.

Chapter Two: Government and Astronomy

1 This thesis is developed in Herschell Hardin, *A Nation Unaware: The Canadian Economic Culture* (Vancouver 1974)
2 See Howard S. Miller, *Dollars for Research: Science and its Patrons in 19th-Century America* (Seattle 1970)
3 A. Vibert Douglas, *Queen's Quarterly*, 78 (winter 1971), 592–601
4 See A.D. Thiessen, *JRASC*, 34 (1940), 308–48, and ibid, 35 (1941), 141–50, 205–24; 36 (1942), 61–5, 457–72.
5 For Lefroy's life, see Carol Whitfield and R.A. Jarrell, *Dictionary of Canadian Biography*, XI. His account of the expedition is given in G.F.G. Stanley, ed., *In Search of the Magnetic North: A Soldier-Surveyor's Letters from the Northwest, 1843–1844* (Toronto 1955).
6 *Canadian Journal*, 1 (1853), 145
7 M.M. Thomson, *The Beginning of the Long Dash* (Toronto 1978), 12ff.
8 University of New Brunswick Archives (UNBA), Observatory Papers, W.B. Jack to W.C. Bond, 10 July 1854
9 Ibid, Jack to G.B. Airy, 19 April 1855
10 Ibid, Jack to Airy, 18 October 1856
11 Ibid
12 Ibid, Airy to Jack, 4 November 1856
13 Ibid, Jack to Airy, 25 November 1856
14 Ibid, Airy to Lords Commissioners of the Admiralty, 22 December 1856
15 For a more detailed account, see Don W. Thomson, *Men and Meridians* (Ottawa 1967), II, 162ff.

16 Province of Canada, *Journals of the Legislative Assembly*, VIII (1849), appendix M.M.M., Earl Grey to Lord Elgin, 26 March 1848
17 Ibid, G.B. Airy to B. Hawes, Colonial Office, 17 July 1848
18 PAC, RG 4, CI, vol. 273, Canada East, Provincial Secretary's Correspondence, Elgin to Grey, 13 June 1848
19 Ibid, Executive Council to J. Leslie, 2 May 1850
20 See R.A. Jarrell, 'Edward David Ashe,' *Dictionary of Canadian Biography*, XII (forthcoming).
21 PAC, RG 4, CI, vol. 273, E.D. Ashe to T.E. Campbell, 18 November 1850
22 Ibid, Ashe to Leslie, 2 December 1850
23 Ibid, Register, vol. 772, entry 1252
24 Province of Canada, *General Index to Journals of the Legislative Assembly, 1852–66*, Supply: Grant to Observatories, 864
25 'Report of Condition of the Quebec Observatory, 1855,' *Journal* (1856), vol. 14, appendix 53
26 Gen Estcourt's commission's value was slightly different from Bayfield's though done in much the same way.
27 Geological Survey of Canada, *Report of Progress for the Years 1853–56* (Toronto 1857), 55
28 Harvard University Archives (HUA), V630.2, Bond Papers, E.D. Ashe to W.C. Bond, 17 March 1857
29 Ashe's results were published in the Geological Survey of Canada, *Reports of Progress, 1858*; a summary appeared in the *Canadian Journal* in 1859.
30 HUA, V630.2, Ashe to Bond, 19 May 1857
31 Ibid, Bond to Ashe, 25 May 1857
32 Ibid, Ashe to Bond, 28 October 1857
33 UNBA, Observatory Papers, Bond to W.B. Jack, 15 April 1857
34 Ibid, G.B. Airy to R.B. Osborne, Admiralty, 30 May 1857
35 His correspondence is in PAC, RG 93D, vol. 82.
36 'Report of the Condition of the Quebec Observatory, 1855'
37 PAC, RG 4, CI, vol. 782, entry 1039, Canada East, Provincial Secretary's Correspondence Register
38 Archives du Séminaire de Québec (ASQ) Université, No. AS, Ashe to T. Roy, 4 February 1858
39 HUA, V630.2, G.P. Bond to Ashe, 5 May 1860
40 E.D. Ashe, *TLHSQ*, 4:2 (1861), 1–16
41 J.E. Kennedy, *JRASC*, 70 (April 1976), 74–6
42 PAC, RG 4, CI, vol. 796, entry 395
43 Canada, Department of Marine and Fisheries, *Annual Report* (1874), 315
44 E.D. Ashe, *TLHSQ*, n.s., part 5 (1866–7), 5–14

45 Ashe, ibid., part 6 (1869), 41–4; the article also appeared under the same title, in *MNRAS*, 26 (1869) 61–2.

46 *MNRAS* (29 January 1869), 275–7

47 PAC, RG93D, vol. 82, Quebec Observatory Correspondence, H. Vogel to E.D. Ashe, 7 June 1869

48 The reports of these observers appear in 'The Canadian Eclipse Party, 1869,' *TLHSQ*, n.s., part 7 (1870), 85–110

49 Cited by Ashe in Marine and Fisheries, *Annual Report* (1871), 138

50 PAC, RG 42, II, B8, vol. 1022, Marine and Fisheries Letterbooks, Hon Peter Mitchell to Privy Council, 28 October 1870

51 Lord Lindsay to Ashe, n.d. (1870), printed in Marine and Fisheries, *Annual Report* (1871), 138–9

52 Ibid, 139

53 Marine and Fisheries, *Annual Report* (1872), 188

54 Ashe to Andrew Elvins, 16 March 1871; printed in C.A. Chant, *JRASC*, 13 (March 1919), 110–11

55 PAC, MG 24, K24, Queen's University Observatory Papers, Thomas Devine to James Williamson, 9 October 1856; Williamson's article appeared in *Canadian Journal*, 3 (1854–5), 82

56 'Queen's College Observatory,' *Queen's College Journal* (26 November 1881), 1

57 Ibid, 4

58 J. Williamson, 'Donati's Comet,' *Canadian Journal*, n.s. 3, 18 (Nov. 1858), 486–8

59 Queen's University Archives (QUA), Kingston Observatory Papers, 1861–91, Deed (dated 19 January 1861)

60 HUA, V630, W. Leitch to G.P. Bond, 16 May 1861

61 Ibid, Bond to Williamson, 8 June 1861

62 Ibid, Williamson to Bond, 19 July 1861

63 QUA, Observatory Papers, Leitch to Williamson, 16 October 1861

64 Ibid, Petition of the Board of Visitors, February 1862

65 Ibid, E.A. Meredith to Williamson, 12 November 1862

66 Ibid, George Romanes to Williamson, 4 May 1863

67 Ibid, Board of Visitors, Minutes, 30 June 1863

68 QUA, Queen's University Board of Trustees Records, 2 December 1863. His duties are described as being required to be at the observatory for six hours daily in summer, four in winter.

69 QUA, Collection 2259, Williamson Papers, Series C, Williamson, 'N.F. Dupuis,' n.d.

70 HUA, V630, Williamson to G.P. Bond, 12 May 1864

71 *MNRAS*, 28 (Nov. 1867) 12–13

72 PAC, MG 26 A, Macdonald Papers, 155761–6, Williamson to Sir J.A. Macdonald, 1 January 1868

73 See J.S. Marshall, *Three McGill Weather Observatories* (Montreal, 1968)

74 Charles Smallwood, *Canadian Journal*, n.s. 3 (1858), 281–92

75 McGill University Archives (MUA), Board of Governors Minute Book, I, 192

76 Ibid, 262

77 PAC, RG 4, CI, vol. 786, entry 194, Canada East, Provincial Secretary's Correspondence Register

78 MUA, Board Minute Book, I, 347

79 PAC, RG 4, CI, vol. 792, entry 2231

80 MUA, Board Minute Book, I, 356

81 Ibid., II, 92

82 PAC, RG 42, II, B8, Marine and Fisheries Letterbook, Hon Peter Mitchell to Viscount Monck, 15 September 1868

83 Marine and Fisheries, *Annual Report* (1869), 22

84 PAC RG 42, II, B8, Hon Albert Smith to Privy Council, 29 April 1874

85 Ibid, Mitchell to Monck, 16 September 1868

86 Marine and Fisheries, *Annual Report* (1874), 308

87 *Queen's College Journal* (26 November 1881), 6

88 McGill University, *Annual Report* (Montreal 1879), 6–7

89 The appropriation was made on 18 May 1882. There was no debate. Reports of the preparations for the transit are given in *TRSC* I (1882), sec. III, 83–98, and Helen S. Hogg, *JRASC*, 76 (Dec. 1982), 362–70.

90 PAC, MG 26 A, Macdonald Papers, Mungo Turnbull to Sir J.A. Macdonald, 15 July 1882

91 ASQ, Université 83, No. 29, W.A. Ashe to T.-E. Hamel, 5 December 1882

92 A.J. Meadows, *Greenwich Observatory: Recent History (1836–1975)* (London 1975), 45–6

93 Montreal, *Witness*, 13 August 1904; see also Nancy Bignell, *McGill News* (Summer 1962), 16–22

94 W.A. Rogers and C.H. McLeod, *TRSC*, 3 (1885), sec. III, 111–37

95 C.H. McLeod, ibid, 8 (1890), sec. III, 43–7; 10 (1892), sec. III, 29–32

96 PAC, MG 26 A, Macdonald Papers, 175493–4, Duttan to Macdonald, 20 September 1881

97 ASQ, *Journal du Séminaire*, 4 (1894), 351

98 Quebec, *L'Electeur*, 16 June 1894

99 PAC, MG 26 G, Laurier Papers, 18030–1, A.-P. Roy à l'hon. Wilfrid Laurier, 30 septembre 1897

100 Ibid, 18029, l'hon. F. Langelier à Roy, 19 mai 1897

101 Ibid, 18031, Roy à Laurier, 30 septembre 1897

102 See J.S. Plaskett, *JRASC*, 10 (July–Aug. 1916), 267–74
103 'Edouard Gaston Daniel Deville,' *PRSC* (1925), viii–xi
104 E.G.D. Deville, *Examples of Astronomic and Geodetic Calculations for the Use of Land Surveyors* (Quebec 1878)
105 Klotz's description of his work is in Department of the Interior, *Annual Report* (1888), part 11, 8–16
106 For the chronology of the foundation of the observatory, see Klotz *JRASC*, 13 (Jan. 1919), 1–15
107 PAC, RG 88, vol. 474, Thomson Papers, Edgar Dewdney to Privy Council, 23 June 1890
108 PAC, MG 27 II, D 15, vol. 45, Sifton Papers, W.F. King to Hon Clifford Sifton, 15 November 1898
109 Ibid,, vol. 103, King to Sifton, 14 June 1901
110 PAC, RG 15, vol. 681, Department of the Interior Records, King to T.G. Rothwell, Acting Deputy Minister, 20 March 1903
111 PAC, RG 88, vol. 473, J.S. Plaskett to King, 31 October 1902
112 PAC, MG 27 II, D15, vol. 250, Sifton to King, 19 February 1903

Chapter Three: Students and Public

1 There is no comprehensive study of scientific education in Canada. A basic survey for universities in R.S. Harris, *A History of Higher Education in Canada 1663–1900* (Toronto 1976).
2 Science education in French Canada has had little attention. For the classical colleges, see Claude Galarneau, *Les collèges classiques au Canada français* (Montreal 1978); for the Séminaire de Québec, see Honorius Provost, *Le Séminaire de Québec: documents et biographies* (Québec, 1964).
3 Archives du Séminaire de Québec (ASQ) Séminaire 92, no. 4
4 Ibid, no. 6
5 Séminaire 152, no. 85 and 85A
6 ASQ, *L'Abeille*, 24 May 1853
7 Séminaire 92, no. 8
8 M-1014, tablette 32
9 See Auguste Gosselin, *MSRC*, sec. 1 (1907), 127–72
10 ASQ, Polygraphie 42, no. 14, Holmes to Airy, 4 May 1837
11 Ibid, no. 14A
12 Quoted in Provost, *Séminaire de Québec*, 328
13 T.-E. Hamel, *PRSC*, 5 (1887)
14 ASQ, *Journal du Séminaire* 4 (18 October 1892), 209
15 See Harris, *History of Higher Education*, chap. 3.
16 For the University of New Brunswick, consult W.O. Raymond, *The Genesis*

of the University of New Brunswick (Saint John 1919), and Richard A. Jarrell, *Acadiensis* (spring 1973), 55–79.

17 Roy Bishop, *JRASC* 72 (June 1978), 138–48; see also William Calnen, ibid., 74 (April 1980), 57–63.

18 King's College, *Calendar* (1863), 29

19 See J.E. Kennedy: *JRASC*, 49 (Sept.–Oct. 1955), 181–8; ibid., 70 (Oct. 1976), 238–46

20 UNBA, UA 2163, J. Robb and W.B. Jack to Chancellor and Council of King's College, 22 February 1847

21 J. Robb, *Oration Delivered at the Encaenia in King's College* (Fredericton, 1839), 12

22 University of New Brunswick Archives (UNBA), UA 2136, Minute Book of the College Council, 23 March 1847, 315

23 UNBA, UA 2162, J. Robb to W. Colebrooke, 12 November 1847

24 Ibid, E. Sabine to Lord Clarence Paget, 13 January 1847

25 UNBA, UA 655, Minute Book, 27 March 1847, 317

26 Ibid, 5 April 1848, 318

27 Ibid, 12 October 1848, 347

28 UNBA, UA 3003, Report of the Committee for Erection of Observatory, 19 March 1851

29 UNBA, Observatory Papers, W.B. Jack to G.B. Airy, June 1854

30 UNBA, UA 3024, W.B. Jack to Bishop of Fredericton, Hon J.A. Street, and Mr Justice Wilmot, 10 February 1853

31 Ibid,

32 See J.E. Kennedy, *Queen's Review* (December 1954), 246–7, 264

33 D.D. Calvin, *Queen's University at Kingston* (Kingston 1941), 72. On Williamson, see R.A. Jarrell, 'James Williamson,' *Dictionary of Canadian Biography*, XII (forthcoming).

34 Queen's University Archives (QUA), University Letters, TC 2

35 QUA, James Williamson Papers

36 Williamson's account of Dupuis's life is included in his papers in QUA; see also J. and R.W. Cumberland, *Queen's Quarterly* (March 1927), 6ff

37 See W.S. Wallace, *A History of the University of Toronto, 1827–1927* (Toronto 1927)

38 Quoted in Harris, *History of Higher Education*, 54

39 This is outlined in Loudon's unpublished memoirs in the University of Toronto Archives (UTA).

40 Egerton Ryerson, *On the Course of Collegiate Education* (Toronto 1842)

41 Nathanial Burwash, *The History of Victoria College* (Toronto 1927), 242

42 Private communication from P. Mozel, 21 May 1979

43 The earliest history is recounted by James Douglas jr., *TLHSQ*, n.s. pt. 4

(1866), 10–13. A general account is the *Centenary Volume of the Literary and Historical Society of Quebec* (Quebec 1924).

44 J. Fletcher, *TLHSQ*, 4, pt. 1 (1843), 1–10

45 V. Daintry, ibid., 11–15

46 W.A. Ashe, ibid., n.s. 19 (1889)

47 ASQ, *Journal du Seminaire*, 4, 209

48 See J.E. Kennedy, *JRASC*, 66 (April 1972), 83–98

49 Great Britain, Meteorological Office Archives, Historical Letters, (20) / 13E, Riddell to Sabine, 5 March 1840

50 J.B. Cherriman and G. Irving, *Canadian Journal*, 2 (1853–4), 253

51 Their account of the eclipse is in ibid., 3 (1854–5), 177–86

52 Col Baron de Rottenburg, ibid., n.s. 1 (1856), 424–9

53 Rottenburg, ibid., n.s. 2 (1857), 180–5

54 Rottenburg, ibid., n.s. 3, (1858), 293–7

55 T. Henning, ibid., 2 (1853–4), 188–91; ibid., 3 (1854–5), 206–9

56 Peter Broughton, *JRASC*, 74 (April 1980), 76–80

57 For the early history of the Royal Society of Canada, see *The Royal Society of Canada, 1882–1957* (Ottawa 1958).

58 The early history of the RASC is outlined by Albert Watson, *JRASC*, 11 (Feb. 1917), 47–78; an abridged version may be found in James Kemp, *The Centennial of the Toronto Centre* (Toronto 1968). Other valuable articles appear in the RASC volume *Astronomy in Canada, Yesterday, Today and Tomorrow* (Toronto 1967), a special number of the *JRASC*. Articles in that collection will be cited by their *Journal* references. Most of this account is drawn from the RASC archives (originals of much of which are now in PAC).

59 University of Pittsburgh, Industrial Society Archives, Keeler Papers

60 G.E. Lumsden, *TRSC*, ser. 2, 2 (1896), sec. III, 83–90

61 During the 1970s, the Association for the Advancement of Science in Canada appeared, at least partly the result of the science policy debates of the 1960s and early 1970s. The still-struggling AASC has always focused on public awareness of scientific issues and has never been a universal forum for scientific papers.

Chapter Four: The Dominion Observatory

1 Canada, Parliament, Sessional Papers, *Report of the Chief Astronomer* (1905), appendix 3

2 PAC, RG 15, B1a, vol. 268, Department of the Interior Papers, W.F. King to T.G. Rothwell, 25 February 1902.

3 Ibid, Hon. J.W. Fielding to J.A. Smart, 15 March 1902

4 Ibid, vol. 294, J.R. Collins to Rt Hon Wilfrid Laurier, 19 November 1904

5 Ibid, King to Hon Clifford Sifton, 7 December 1904

6 Plaskett's description as found in the *Report of the Chief Astronomer* (1905), appendix 5

7 See A.T. De Lury, *Transactions of the Royal Astronomical Society of Canada*, 1905, 57–69

8 An overview of activities up to 1930 can be found in the 'Activities of the Astronomical Branch, 1930,' PAC, RG 48. For surveying, see Thomson, *Men and Meridians*, vols. 2 and 3; for timekeeping, Thomson, *Beginning of the Long Dash*, 74ff.

9 See F.A. McDiarmid, *JRASC*, 17 (Nov.–Dec. 1923), 357–9

10 For Plaskett's role, see R.A. Jarrell, ibid., 71 (June 1977), 221–33

11 *Report of the Chief Astronomer* (1907), 47–73

12 Ibid, 74

13 Lick Observatory Archives (LOA), Plaskett to W.W. Campbell, 17 March 1906

14 Harvard University Archives (HUA), Pickering Papers, Plaskett to E.C. Pickering, 9 September 1907

15 Ibid, Plaskett to E.C. Pickering, 14 January 1909

16 R.M. Motherwell, *JRASC*, 8 (Sept.–Oct. 1914), 305–6

17 Werry had gone to Saranac Lake, NY, in February 1905, but King had arranged for work to be sent to him. PAC, RG 15, vol. 681, King to W.W. Cory, 19 November 1906

18 *Report of the Chief Astronomer* (1910), 87, 93ff

19 George Ellery Hale Papers (microfilm) (GEHP), (Roll) 95, Plaskett to G.E. Hale, 27 March 1906

20 Ibid, Hale to Plaskett, 28 April 1906

21 Ibid, Plaskett to Hale, 9 May 1906. This was spelled out further in a letter to Hale of 21 July 1910, in which Plaskett notes that his background in radial velocities suited him best for this line of work.

22 W.F. King, *JRASC*, 3 (1909), 115–6

23 GEHP, 95, Hale to Plaskett, 6 January 1910

24 Plaskett's report of the meeting is in *Report of the Chief Astronomer* (1911), 139–43

25 GEHP, 85, Campbell to W.S. Adams, 28 December 1911

26 *Report of the Chief Astronomer* (1911), 150

27 Ibid, 149–50, Plaskett to F. Schlesinger, 26 January 1911

28 GEHP, 95, Plaskett to Adams, 26 July 1910

29 Ibid, Plaskett to Adams, 3 February 1912

30 Ibid, Adams to Plaskett, 19 February 1912

31 Ibid, Plaskett to Adams, 26 November 1914

32 A copy of H.H. Plaskett's circular is in the Hale Papers, 95
33 Ibid, 86, Hale to R.E. De Lury, 26 November 1913
34 H.H. Plaskett, *JRASC*, 8 (Sept.–Oct. 1914), 307–17; his article on the psychological factors in personal equation appeared in ibid, 10 (May–June 1916), 220–34.
35 GEHP, 95, Adams to J.S. Plaskett, 16 October 1914
36 Ibid, Plaskett to Adams, 19 February 1915
37 R.E. De Lury, *JRASC*, 10 (Sept. 1916), 345–57; C.E. St. John and W.S. Adams, ibid., 10 (Dec. 1916), 553–6; De Lury's response, ibid., 11 (Jan. 1917), 22–3
38 GEHP, 95, Plaskett to Adams, 23 December 1916
39 H.H. Plaskett, *JRASC*, 13 (Nov. 1919), 391–402; De Lury, ibid., 13 (Dec. 1919), 449–51
40 *Report of the Chief Astronomer* (1911), 140
41 LOA, Plaskett to Campbell, 7 November 1910
42 HUA, Pickering Papers, Plaskett to Pickering, 18 May 1912
43 LOA, Plaskett to Campbell, 18 May 1912
44 *JRASC*, 7 (Jan.–Feb. 1913), 13, 46–7
45 W.F. King, ibid. (May, June 1913), 216–28. Plaskett's summary of the steps taken are in: ibid., 7 (Nov.–Dec. 1913), 448–55
46 PAC, RG 2, 4, vol. 45 (1913), entry 2581, Canada, Privy Council Register
47 W.E. Harper, *JRASC*, 8 (May–June 1914), 172–9
48 GEHP, 86, W.F. King to G.E. Hale, 6 March 1913; LOA, Plaskett to Campbell, 5 March 1913
49 LOA, Campbell to Plaskett, 28 October 1913
50 Dominion Astrophysical Observatory (DAO), Plaskett to McBride, 5 March 1914. See also J.S. Plaskett, *JRASC*, 8 (May–June 1914), 180–7
51 DAO, W.F. King to J.B. Hunter, Deputy Minister Public Works, 24 October 1914
52 Ibid, F.H. Shepherd to Plaskett, 18 February 1915
53 Ibid, Shepherd to Hon Robert Rogers, 25 February 1915
54 Ibid, Shepherd to G. Gray Donald, 10 March 1915
55 Ibid, Rt Hon Robert Borden to McBride, 26 November 1915 (copy enclosed in letter, McBride to Plaskett, 2 December 1915)
56 Ibid, Hon W.J. Roche to Borden, 25 November 1915 (copy enclosed in McBride letter of 2 December 1915)
57 HUA, Pickering Papers, Pickering to Plaskett, 24 December 1915; Plaskett to Pickering, 27 December 1915
58 Ibid, Plaskett to Pickering, 7 May 1918
59 PAC, RG 88, vol. 473, Plaskett to W.F. King, 19 February 1916
60 A copy of the circular is in the GEHP, roll 95.

61 Ibid, Plaskett to Hale, 15 May 1916
62 LOA, Plaskett to Campbell, 30 May 1916
63 GEHP, 95, Roche to Hale, 21 June 1916
64 PAC, MG 30, CI, O.J. Klotz Diary, vol. 7, 190
65 Ibid, 191
66 PAC, RG 2, vol. 49 (1917), entry 992, Canada, Privy Council Register
67 Ibid, vol. 473, entry 31/992
68 Ibid, vol. 49, entries 2755–6
69 PAC, RG 48, R.M. Stewart, 'Activities of the Astronomical Branch,' G–2
70 F. Henroteau and H. Grouiller, *JRASC*, 19 (Oct. 1925), 201–3
71 Otto Struve and V. Zebergs, *Astronomy of the Twentieth Century* (New York 1962), 73–4
72 Interview with J.A. Pearce, Victoria, BC, August 1973
73 LOA, R.M. Stewart to R.G. Aitken, 10 April 1928
74 Ibid, Stewart to Aitken, 14 May 1929.
75 See R.J. McDiarmid, *JRASC*, 34 (Dec. 1940), 441–3
76 R.E. De Lury, ibid, 33 (Nov. 1939), 345–78

Chapter Five: The Dominion Astrophysical Observatory

1 DAO, Plaskett to Harper, 27 May 1918
2 Ibid, Harper to Plaskett, 12 June 1918
3 Ibid, Plaskett to Harper, 22 October 1918
4 GEHP, 95, Plaskett to Adams, 10 October 1918; Adams to Plaskett, 20 October 1918
5 Ibid, J.C. Kapteyn to Plaskett, 16 March 1918
6 Ibid, Plaskett to Adams, 25 May 1918
7 Ibid, Plaskett to Hale, 31 July 1918
8 Ibid, Plaskett to Adams, 3 October 1918
9 J.S. Plaskett, W. Harper, R. Young, and H. Plaskett, *Pubs. DAO* (1921)
10 DAO, Hale to Plaskett, 27 September 1921
11 DAO, Campbell to Plaskett
12 DAO, Plaskett to W.W. Cory, 6 March 1919. The statement was more peevish than accurate, since Young's tables were precisely the kind of astronomy Klotz excelled in.
13 DAO, Plaskett to W.L. Mackenzie King, 27 September 1920
14 Ibid, Plaskett to Mackenzie King, 27 March 1923; Mackenzie King to Plaskett, 19 April 1923
15 Ibid, Plaskett to Mackenzie King, 3 January 1927
16 Ibid, W.W. Cory to Plaskett, 20 January 1927
17 H.H. Plaskett, *Pubs. DAO*, 1, 30 (1922), 225–84

18 R.K. Young, ibid., 1, 2 (1918), 105–11; an earlier orbit of the star was published in *Pubs. DO*, 3, 3.

19 François Henroteau, *JRASC*, 15 (Feb. 1921), 62–70; (March 1921), 109–19. Cf. Henroteau and J.P. Henderson, *Pubs. DO*, 5, 1 (1922)

20 See J.S. Plaskett, *JRASC*, 16 (Nov. 1922), 284–93

21 J.S. Plaskett, *Pubs. DAO*, 2, 14 (1923), 269–74

22 J.S. Plaskett, ibid., 2, 16 (1924), 287–356

23 R.K. Young and W.E. Harper, ibid., 3, 1, (1924), 3–143; an earlier version, comprising some 1080 stars, was published in the *JRASC*, 18 (Jan.–Feb. 1924), 9–59.

24 Lick Observatory Archives (LOA), J.S. Plaskett to R.G. Aitken, 24 January 1924

25 Ibid, R.G. Aitken to J.S. Plaskett, 29 January 1924

26 For a readable survey of this issue, see R. Berendzen, R. Hart, and D. Seeley, *Man Discovers the Galaxies* (New York 1978).

27 LOA, J.A. Pearce to R.G. Aitken, 8 January 1925; Pearce to W.H. Wright, 30 April 1925

28 J.S. Plaskett and J.A. Pearce, *Pubs. DAO*, 5, 1, (1930), 1–98

29 J.S. Plaskett, *MNRAS*, 88 (March 1928), 395–403

30 J.S. Plaskett and J.A. Pearce, *Pubs. DAO*, 5, 4 (1934), 241–328

31 For Redman, see C.S. Beals, *JRASC*, 70 (Feb. 1976), 34–9. Redman was Beals's brother-in-law.

32 R.O. Redman, *MNRAS*, 90 (May 1930), 690–6; further details are in *Pubs. DAO*, 4, 20 (1930), 325–40 and 6, 5, (1931), 27–48.

33 Berendzen et al., *Man Discovers the Galaxies*, 86–7

34 J.S. Plaskett and J.A. Pearce, *MNRAS*, 90 (Jan. 1930), 243–68. Plaskett also published a popular account in *Popular Astronomy* the following month.

35 J.S. Plaskett and J.A. Pearce, *Pubs. DAO*, 5, 3 (1931), 167–223

36 LOA, R.G. Aitken to Plaskett, 4 January 1926

37 J.L. Locke, *JRASC*, 73 (Dec. 1979), 325–32

38 C.S. Beals, *Pubs. DAO*, 4, 17 (1929), 271–301. Cf. *JRASC*, 24 (July–Aug. 1930), 277–81.

39 C.S. Beals, *Pubs. DAO*, 6, 9 (1933), 95–148

40 LOA, Beals to W.H. Wright, 17 July 1933

41 C.S. Beals, *JRASC*, 60 (Aug. 1966), 157–66

42 C.S. Beals, ibid., 54 (Aug. 1960), 153–6

43 C.A. Chant, ibid., 45 (Jan.-Feb. 1951), 1–3

44 Helen B. Sawyer, *Pubs. DAO*, 6, 14, (1935), 265–84

45 LOA, Aitken to Plaskett, 7 March 1933

46 J.A. Pearce, personal interview, August 1973

47 R.M. Stewart, *JRASC*, 34 (Jan.–Feb. 1940), 233–7

48 R.M. Petrie, *Pubs. DAO*, 7, 21 (1947), 321ff.
49 C.S. Beals, *MNRAS*, 96 (1936), 661–79
50 Andrew McKellar, *Publications of the Astronomical Society of the Pacific*, 52 (1940), 187–92
51 Andrew McKellar, *Pubs. DAO*, 7, 15 (1941), 251–72
52 LOA, W.E. Harper to W.H. Wright, 30 August 1937
53 Ibid, Beals to J.H. Moore, 24 September 1936
54 C.S. Beals and A. McKellar, *JRASC*, 32 (Oct. 1938), 369–75
55 J.A. Pearce, ibid., 36 (July–Aug. 1942), 273–87; for 1942–5, see Pearce, ibid., 40 (April 1946), 139–55.
56 Donald Osterbrook, *Journal for the History of Astronomy*, 15, 2 (June 1984), 81–127; 15, 3 (Oct. 1984), 151–76

Chapter Six: Universities and Associations

1 See appendix B.
2 For a summary of the work at Toronto, see J.F. Heard and H.S. Hogg, *JRASC*, 61 (Oct. 1967), 257–76.
3 Lick Observatory Archives (LOA), Chant to Campbell, 1 October 1906
4 Ibid, Chant to Campbell, 8 June 1908
5 Ibid, Chant to Campbell, 29 October 1908
6 Ibid, Chant to Campbell, 10 March 1910
7 Ibid, Chant to Campbell, 19 March 1912
8 Ibid, Chant to Campbell, 19 February 1912
9 C.A. Chant, *JRASC*, 12 (Sept. 1918), 339–49
10 C.A. Chant and R.K. Young, *Pubs. DAO*, 2, 15 (1923), 275–85
11 Chant's account of the observatory was one portion of his unpublished autobiography, now in the University of Toronto Archives. The chapter devoted to the origins of the DDO was published as *Astronomy in the University of Toronto: The David Dunlap Observatory* (Toronto 1954). Young's technical account appeared in instalments in *Engineering*, 9 and 30 March, 20 April 1934.
12 George Ellery Hale Papers (microfilm) (GEHP), 85 Chant to Hale, 1 May 1920; Hale to Chant, 19 May 1920
13 R.K. Young, *JRASC*, 24 (Jan. 1930), 17–33
14 R.K. Young, ibid., 36 (April 1942), 142–4
15 University of Toronto, *President's Report* (1938), 162.
16 Ibid, (1941), 153
17 Ibid, (1947), 143
18 University of Toronto Archives (UTA), A74–027, Chant Correspondence, Gillson to Chant, 4 November 1923; Gillson to Chant, 7 February 1924

19 Published in *JRASC*, 20, (Oct. 1926), 265–302
20 F. Henroteau and A.V. Douglas, *Pubs. DO*, 9, 7 (1929), 163–77
21 J.S. Foster and A.V. Douglas, *Physical Review*, 44 (1933), 325; *Nature*, 134 (1934), 417–18
22 A.B. Underhill and W. Petrie, *JRASC*, 38 (Nov. 1944), 385–94
23 Otto Struve and V. Zebergs, *Astronomy of the Twentieth Century* (New York 1962), 494
24 Details of the history of Queen's program are given by Douglas, *JRASC*, 52 (April 1958), 82–6
25 H.R. Kingston, ibid., 34 (March 1940), 95–7
26 M.M. Thomson, personal interview, May 1973
27 H.L. Welsh, *JRASC*, 66 (Aug. 1972), 183–8
28 E.S. Keeping, *A Short History of the Mathematics Department* (Edmonton 1971)
29 A brief sketch of the history of the RASC is: Ruth J. Northcott, *JRASC*, 61 (Oct. 1967), 218–25.
30 GEHP, 85, Hale to Chant, 22 October 1908
31 Ibid, Chant to Hale, 2 February 1909
32 *JRASC*, 1 (Jan.–Feb. 1907), 1
33 A brief history of the Edmonton Centre has been prepared by E.S. Keeping: 'The Earlier Years of the Edmonton Centre,' unpublished typescript, RASC Archives
34 LOA, Chant to Aitken, 12 October 1934
35 Archives du Séminaire de Québec (ASQ), Université 286, no. 45, 'Rapport du conseil du Séminaire de Québec' (c. May 1941)
36 ASQ, Université 282, no. 79, Nadeau à Mgr C. Roy, 23 May 1941
37 Ibid
38 See R.A. Jarrell, *JRASC*, 73 (Dec. 1979), 358–69
39 LOA, Chant to Aitken, 27 September 1926
40 C.A. Chant, *JRASC*, 5 (Sept.–Oct. 1911), 327–38
41 C.S. Beals, personal interview, May 1973
42 J.S. Plaskett, *JRASC* 7 (Nov.–Dec. 1913), 420–37
43 GEHP, 29, Plaskett to Hale, 3 July 1919
44 Ibid, Hale to Plaskett, 9 July 1919
45 See *JRASC*, 14 (June 1920), 219–20
46 A.V. Douglas, *The Life of Arthur Stanley Eddington* (London 1956)

Chapter Seven: New Centres, New Directions

1 C.S. Beals, personal interview, May 1973
2 Obtaining student assistants and new staff members was not as easy in the

late 1940s as it was earlier, since post-war students were more likely to be attracted by astrophysical research and not by what Beals admitted to Frank Hogg was the 'donkey work' of astronomy.

3 C.A. Chant, *JRASC*, 7 (May–June 1913), 145–215

4 For a survey of meteor astronomy in Canada, see P.M. Millman and D.W.R. McKinley, ibid, 61 (Oct. 1967), 277–94.

5 P.M. Millman, ibid., 53 (Feb. 1959), 15–33

6 For the history of the NRC, see Wilfrid Eggleston, *National Research in Canada* (Montreal 1978), and W.E.K. Middleton, *Physics at the National Research Council* (Waterloo 1980).

7 P.M. Millman and D.W.R. McKinley, *JRASC*, 42 (May–June 1948), 121–30

8 D.W.R. McKinley, *Astrophysical Journal (Ap. J.)*, 113 (1951), 225–67

9 P.M. Millman, *JRASC*, 44 (Nov.–Dec. 1950), 209–20

10 V.B. Meen, ibid, 44 (Sept.–Oct. 1950), 169–80

11 C.S. Beals and I. Halliday, ibid, 61 (Oct. 1967), 295–313

12 I. Halliday, A. Blackwell, and A. Griffin, ibid, 72 (Feb. 1978), 15–39

13 For surveys of meteor astronomy with overviews of Canadian contributions, see P.M. Millman, in I. Halliday and B. McIntosh, eds., *Solid Particles in the Solar System*, 121–8; and Millman and D.W.R. McKinley, in B. Middlehurst and G. Kuiper, eds., *The Solar System*, IV, 674–773

14 For background, consult A.E. Covington, *JRASC*, 61 (Oct. 1967), 314–23; see also Covington, *Historical Background for the Absolute Calibration of Solar Flux Density at 2800 MHz* (Ottawa 1979).

15 A.E. Covington, *Journal of Terrestrial Magnetism*, 52 (1947), 339–41, and in *Journal of Geophysical Research*, 55 (1950), 33–7

16 A.E. Covington, *Nature*, 159 (1947), 405–6

17 A.E. Covington and W.J. Medd, *JRASC*, 43 (May–June 1949), 106–10

18 A.E. Covington and N. Broten, *Ap. J.*, 119 (1954), 569–89

19 Private communication, A.E. Covington, March 1981

20 PAC, RG 48, C.S. Beals Papers, 610–11

21 J.L. Locke, *JRASC*, 61 (Oct. 1967), 328. A description of the radio telescope is given in Locke, J. Galt and C. Costain, *Pubs. DO*, 25, 7, (1965), 77–82

22 M.B. Bell, A.E. Covington, and W.A.G. Kennedy, *Solar Physics*, 28 (1978), 126–36

23 N. Broten *et al*, *Nature*, 215 (1967), 38; see also N. Broten et al, *Science* 156 (1967), 1592–3

24 L. Avery et al, *Ap. J. Letters*, 205 (1976), L173–5, and N. Broten et al, ibid., 223 (1978), L105–7

25 J.D. Lacey, *JRASC*, 62 (Dec. 1968), 378–80

26 V. Gaizauskas, ibid., 70 (Feb. 1976), 1–22

27 R.M. Petrie and J.A. Pearce, *Pubs. DAO*, 12, 1 (1962), 1–90

28 R.M. Petrie, ibid., 9, 8 (1953), 251–67
29 R.M. Petrie and J.K. Petrie, ibid., 13, 9 (1968), 253–72. A popular account is given by J.K. Petrie, *JRASC*, 64 (June 1970), 163–72
30 E.H. Richardson, ibid., 62 (Dec. 1968), 313–30
31 E.H. Richardson, G. Brealey, and R. Dancey, *Pubs. DAO*, 14, 1 (1971), 1–15
32 See K.O. Wright, *JRASC*, 69 (Oct. 1975), 205–11
33 A.H. Batten, *Pubs. DAO*, 13, 8 (1967), 119–251
34 K.O. Wright et al, *JRASC*, 57 (April 1963), 66–72
35 R.M. Petrie, ibid., 57 (Aug. 1963), 145.
36 *Report of Preliminary Studies, Mt. Kobau National Observatory (Final Draft, March 31, 1977)*; copy in PAC, RG 48, vol. 2, C.S. Beals Papers
37 The policy-forming system at the time is described in G. Bruce Doern, *Science and Politics in Canada* (Montreal 1972).
38 C.S. Beals, *JRASC*, 69 (Feb. 1975), 35–6
39 Science Secretariat Working Group on Astronomy (mimeograph report, 1968)
40 This was reported to the RASC by G. Odgers and K.O. Wright, *JRASC*, 62 (Dec. 1968), 392–4.
41 A survey of astronomical personnel and instrumentation is: Associate Committee on Astronomy, *Canadian Facilities for Research in Astronomy* (Ottawa, 1977)
42 See Malcolm M. Thomson, *The Beginning of the Long Dash: A History of Timekeeping in Canada* (Toronto 1978), chap. 5.
43 J.L. Locke, *JRASC*, 71 (Feb. 1977), 9–20
44 Associate Committee on Astronomy, *The Future of Ground and Space Based Astronomy in Canada* (Ottawa 1974)
45 Working Group on Space Astronomy, *A Report Submitted to the Ad Hoc Committee on Canadian Scientific Experiments for the Space Shuttle / Spacelab Programme* (Ottawa 1977); see also the ad hoc committee's *Report Submitted to the Space Science Coordination Office, National Research Council* (Ottawa 1977)
46 René Racine, *JRASC*, 70 (June 1976), 138–42, and ibid, 72 (Dec. 1978), 324–34
47 George Harrower, ibid., 54 (April 1960), 82–3
48 William Wehlau, ibid., 64 (Feb. 1970), 1–4
49 The files of the astronomy department are at the University of Toronto Archives (UTA), A-74-027 and A-75-0047.
50 R.C. Roeder and P.P. Kronberg, *JRASC*, 64 (Oct. 1970), 315–18
51 J.R. Percy, ibid., 72 (Oct. 1978), 296–8
52 J.R. Percy, ibid., 71 (June 1977), 264–74
53 One measure of this respect is the IAU's naming of lunar features – such as

the crater Chant – and, recently, of asteroids for H.S. Hogg, J.S. and H.H. Plaskett, J.F. Heard, P.M. Millman, A.B. Underhill, and J. Climenhaga.

Conclusion

1 The argument for Canada's imperial links is made by Vittorio de Vecchi, 'Science and Government in Nineteenth-Century Canada,' PH D dissertation, University of Toronto, 1978; but see also R.A. Jarrell, *TRSC*, ser. IV, 20 (1982), 533–47.

Bibliography

Abbreviations

Ap. J. *Astrophysical Journal*
JRASC *Journal of the Royal Astronomical Society of Canada*
MNRAS *Monthly Notices of the Royal Astronomical Society*
MSRC *Mémoires de la Société royale du Canada*
PAC *Public Archives of Canada, Ottawa*
Phil. Trans. *Philosophical Transactions, Royal Society of London*
PRSC *Proceedings, Royal Society of Canada*
Pubs. DAO *Publications, Dominion Astrophysical Observatory*
Pubs. DO *Publications, Dominion Observatory*
THLSQ *Transactions, Literary and Historical Society of Quebec*
TRSC *Transactions, Royal Society of Canada*

Archival Sources

Dominion Astrophysical Observatory, Victoria, BC
  Correspondence Files
Harvard University Archives, Cambridge, Mass
  W.C. Bond Papers
  E.C. Pickering Papers
Lick Observatory Archives, Santa Cruz, Calif.
  Observatory Correspondence Files
McGill University Archives, Montreal
  C. Kirkland McLeod Papers
  McGill College Annual Reports
  McGill College Board of Governors Minute Books

Public Archives of Canada, Ottawa
  MG 12, DI2, T28, V. 14–15. Great Britain, Treasury, Out Letters, 1816–17
  MG 24, F28, V. 2. Henry Bayfield Papers
  MG 24, K24. Queen's University Observatory Papers
  MG 26, A. Sir John A. Macdonald Papers
  MG 26, G. Sir Wilfrid Laurier Papers
  MG 27, II, DI5, V. 45. Clifford Sifton Papers
  MG 30, cl. Otto Klotz Papers
  RG 2, 4. Privy Council Registers
  RG 4, cl. Canada East, Provincial Secretary's Correspondence
  RG 5, cl. Canada West, Provincial Secretary's Correspondence
  RG 15, B, V. 681. W.F. King's Personnel File
  RG 15, B. Department of the Interior Records
  RG 42 II, B8, V. 1022–34. Department of Marine and Fisheries Letterbooks
  RG 48. Dominion Observatory, General Administrative Records
  RG 48. C.S. Beals Papers
  RG 88, V. 473–4. Thomson Papers
  RG 93D, V. 82. Quebec Observatory Correspondence
Queen's University Archives, Kingston, Ontario
  Board of Trustees Records
  Kingston Observatory Papers
  James Williamson Papers
Royal Astronomical Society of Canada Archives, Toronto
  E.S. Keeping, 'Early History of Edmonton Centre' (typescript)
  RASC Records
  Toronto Astronomical and Physical Society Records
Séminaire de Québec, Archives
  Journal du Séminaire
  MI014 (cahier)
  Polygraph 42 (lettres)
  Séminaire 92 (lettres)
  Séminaire 152 (lettres)
  Université 83 (lettres)
  Université 101 (lettres)
University of New Brunswick Archives, Fredericton, NB
  King's College Council Minute Books
  Observatory Papers
University of Pittsburgh, Industrial Society Archives
  J.E. Keeler Papers

University of Toronto Archives
A74–027 David Dunlap Observatory Papers
A75–0047 David Dunlap Observatory Papers

Documentary Sources and Series

American Institute of Physics, *George Ellery Hale Papers*
Calendars, universities of Acadia, Alberta, British Columbia, Dalhousie, King's, Laval, Manitoba, McGill, Montréal, Mt Allison, New Brunswick, Queen's, Saskatchewan, Toronto, Waterloo, Western Ontario, York
Canada, Department of Marine and Fisheries, *Annual Report*
Canada, Department of the Interior, *Annual Report* and *Report of the Chief Astronomer*
Canada, Parliament, House of Commons, *Debates*
Canada, Parliament, *Sessional Papers*
Canada (Province), Geological Survey, *Report of Progress*
Canada (Province), Legislative Assembly, *Journals*
David Dunlap Observatory, *Publications*
Dominion Astrophysical Observatory, *Publications*
Dominion Observatory, *Publications*
Toronto Mechanics' Institute, *Annual Report*
University of Toronto, *President's Report*

Printed Sources

Ashe, E.D. 'The Canadian Eclipse Party, 1869.' *TLHSQ*, n.s., part 7 (1870), 85–110
– 'The Late Eclipse – Journal of a Voyage from New York to Labrador.' *TLHSQ*, 4:2 (1861), 1–16
– 'On the Longitude of Some of the Principal Places in Canada, as Determined by Electric Telegraph in the Years 1856–7.' *Report of Progress of the Geological Survey of Canada 1857*, 231–40, Toronto 1858
– 'On Solar Spots.' *TLHSQ*, n.s., part 5 (1866–7), 5–14
– 'On the Physical Constitution of the Sun.' *TLHSQ*, n.s., part 6 (1869), 41–4
– 'On the Physical Constitution of the Sun.' *MNRAS*, 26 (1869), 61–2
Ashe, William A. 'An Elementary Discussion of the Nebular Hypothesis.' *TLHSQ*, n.s. 19 (1889)
Associate Committee on Astronomy, NRC. *Canadian Facilities for Research in Astronomy*. Ottawa 1977
– *The Future of Ground and Space Based Astronomy in Canada*. Ottawa 1974

'Astronomy in Canada – Canadian Astronomers Report XXI.' *JRASC*, 54
(April 1960), 79–87; (June 1960), 126–39

Audet, Louis-Philippe. *Histoire de l'enseignement au Québec.* 2 vols. Montreal 1971

– 'Hydrographes du roi et cours d'hydrographie au collège de Québec.' *Cahiers des Dix*, 35 (1970), 13–37

Avery, L., et al. 'Detection of the Heavy Interstellar Molecule Cyanodiacetylene.' *Ap. J. Letters*, 205 (1976), L173–5

Bailey, A.G., ed. *The University of New Brunswick Memorial Volume.* Fredericton 1950

Batten, Alan H. 'Of Stars and the Galaxy.' *JRASC*, 61 (Oct. 1967), 241–56

– 'Sixth Catalogue of the Orbital Elements of Spectroscopic Binary Systems.' *Pubs. DAO*, 13, 8 (1967), 119–251

Beals, C.S. 'Andrew McKellar, 1910–1960.' *JRASC*, 54 (Aug. 1960), 153–6

– 'Early Days at the Dominion Astrophysical Observatory.' *JRASC*, 62 (Dec. 1968), 298–312

– 'John Stanley Plaskett.' *JRASC*, 35 (Dec. 1941), 401–7

– 'The Nature of Absorbing Material within the Galaxy and Its Influence on Estimates of Galactic Dimensions.' *JRASC*, 39 (Nov. 1945), 329–46

– 'On the Interpretation of Interstellar Lines.' *MNRAS*, 96 (1936), 661–79

– 'On the Nature of Wolf-Rayet Emission.' *MNRAS*, 90 (1929), 202–12

– 'On the Physical Characteristics of the Wolf Rayet Stars and Their Relation to Other Objects of Early Type.' *JRASC*, 34 (May–June 1940), 169–97

– 'Ralph Emerson De Lury.' *MNRAS*, 117 (1957), 251–2

– 'Robert Methven Petrie, 1906–1966.' *JRASC*, 60 (Aug. 1966), 157–66

– 'Roderick Oliver Redman, FRS, 1905–1975.' *JRASC*, 70 (Feb. 1976), 34–9

– 'Spectrophotometric Studies of Wolf Rayet Stars and Novae.' *Pubs. DAO* 6, 9 (1933), 95–148

– 'William Elgin van Steenburgh, 1899–1974.' *JRASC*, 69 (Feb. 1975), 35–6

– 'The Wolf Rayet Stars.' *Pubs. DAO*, 4, 17 (1929), 271–301

– 'The Wolf Rayet Stars.' *JRASC*, 24 (July–Aug. 1930), 277–81

Beals, C.S., G.M. Ferguson, and A. Landau. 'A Search for Analogies between Lunar and Terrestrial Topography on Photographs of the Canadian Shield.' *JRASC*, 50 (Sept.–Oct. 1956), 203–11; (Nov.–Dec. 1956), 250–61

Beals, C.S., and I. Halliday. 'Impact Craters of the Earth and Moon.' *JRASC*, 61 (Oct. 1961), 295–313

Beals, C.S., and A. McKellar. 'On the Use of Aluminum-on-Glass Gratings in the Victoria Stellar Spectrograph.' *JRASC*, 32 (Oct. 1938), 369–75

Beattie, Brian. 'The 6-inch Cooke Refractor in Toronto.' *JRASC*, 76 (April 1982), 109–28

Bell, M.B., A.E. Covington, and W.A.G. Kennedy. 'Polarization of Interferometer for 2800 MHz Solar Noise Studies with a 0.5' Fan Beam.' *Solar Physics*, 28 (1978), 123–36

Berendzen, R., R. Hart, and D. Seeley. *Man Discovers the Galaxies*. New York 1976

Berger, Carl. *The Writing of Canadian History*. 2nd ed. Toronto 1986

Bignell, Nancy. 'Official Time Signal: 100 Years.' *McGill News* (summer 1962), 16–22

Bishop, Roy. 'An Eighteenth-Century Nova Scotia Observatory.' *JRASC*, 71 (Dec. 1977), 425–42

– 'Joseph Everett and the King's Observatory.' *JRASC*, 72 (June 1978), 138–48

Bouchette, Joseph (fils). *Tables Showing the Difference of Longitude in Time*. Toronto 1857

Brooks, R.C. 'M. de Chabert and the 1750 Louisbourg Observatory.' *JRASC*, 73 (Dec. 1979), 333–48

Broten, Neil, et al. 'The Detection of HC_9N in Interstellar Space.' *Ap. J. Letters*, 223 (1978), L105–7

– 'Long Base Interferometry: A New Technique.' *Science*, 156 (1967), 1592–3

– 'Observations of Quasars using Interferometer Baselines up to 3,074 km.' *Nature*, 215 (1967), 38

Broughton, Peter. 'Astronomy in Seventeenth-Century Canada.' *JRASC*, 75 (Aug. 1981), 175–208

– 'Henry Hayden, 1784–1862.' *JRASC*, 74 (April 1980), 76–80

Brydon, H. Boyd. 'Popularizing Astronomy.' *JRASC*, 29 (Feb. 1935), 41–8

Burland, Miriam. 'Robert Meldrum Stewart. December 15, 1878–September 2, 1954.' *JRASC*, 49 (March–April 1955), 64–8

Burwash, Nathaniel. *The History of Victoria College*. Toronto 1927

Calnen, William J. 'Astronomy at King's College, Windsor, Nova Scotia.' *JRASC*, 74 (April 1980), 57–63

Calvin, D.D. *Queen's University at Kingston*. Kingston 1941

Cassini, J.-D. *A Voyage to Newfoundland and Sallee, to Make Experiments on Mr. Le Roy's Timekeepers*. London 1778

Chabert de Cogolin, J.-B., marquis de. *Voyage fait par ordre du Roi en 1750 et 1751, dans l'Amérique septentrionale pour rectifier les cartes de côtes de l'Acadie, de l'Isle Royale & de l'Isle de Terre-Neuve, et pour en fixer les principaux points par des observations astronomiques*. Paris 1753

Chant, C.A. 'Andrew Elvins (1823–1918).' *JRASC*, 13 (March 1919), 98–121

– 'The Astronomical and Astrophysical Society of America Ottawa Meeting.' *JRASC*, 5 (Sept.–Oct. 1911), 327–38

– *Astronomy in the University of Toronto: The David Dunlap Observatory*. Toronto 1954

- 'An Extraordinary Meteoric Display.' *JRASC*, 7 (May–June 1913), 145–215
- 'The Fiftieth Anniversary of the Royal Astronomical Society of Canada.' *JRASC*, 34 (Sept. 1940), 273–307
- 'Frank Scott Hogg, 1904–1951.' *JRASC*, 45 (Jan.–Feb. 1951), 1–3
- 'John S. Plaskett at the University of Toronto.' *JRASC*, 35 (Dec. 1941), 412–14
- *Our Wonderful Universe*. London 1928
- 'The Solar Eclipse of June 8, 1918; Observations at Matheson, Colorado.' *JRASC*, 12 (Sept. 1918), 339–49
Chant, C.A., and R.K. Young. 'Evidence of the Bending of the Rays of Light on Passing the Sun, Obtained by the Canadian Expedition to Observe the Australian Eclipse.' *Pubs. DAO*, 2, 15 (1923), 275–85
Cherriman, J.B. 'Solar Eclipse, May 26th, 1854.' *Canadian Journal*, 3 (1854–5), 177–86
Cherriman, J.B., and G. Irving. 'Eclipse of the Sun, May 26th, 1854.' *Canadian Journal*, 2 (1853–4), 253
Clark, A.L. *The First Fifty Years: A History of the Science Faculty at Queen's 1893–1943*. Kingston 1944
Cook, James. 'An Observation of an Eclipse of the Sun at the Isle of Newfoundland, August 5, 1766.' *Phil. Trans.*, 57 (1767), 215
Covington, A.E. 'The Development of Solar Microwave Radio Astronomy in Canada.' *JRASC*, 61 (Oct. 1967), 214–23
- *Historical Background for the 1970 Absolute Calibration of Solar Flux Density at 2800 MHz*. NRC Report 17686. Ottawa 1979
- 'Microwave Sky Noise.' *Journal of Geophysical Research*, 55 (1950), 33–7
- 'Microwave Sky Noise.' *Journal of Terrestrial Magnetism*, 52 (1947), 339–41
- 'Microwave Solar Noise Observations during the Partial Eclipse of November 23, 1946.' *Nature*, 159 (1947), 405–6
- 'Solar Radio Astronomy.' *JRASC* 51 (Oct. 1957), 298–307
- 'A 10.5 Year Period for the Slowly Varying Component of the Solar Radio Flux.' *JRASC*, 68 (Feb. 1974), 31–5
Covington, A.E., and N. Broten. 'Brightness of the Solar Disk at a Wave Length of 10.3 cm.' *Ap. J.*, 119 (1954), 569–89
Covington, A.E., and J.L. Locke. '2,700 MC/s Solar Patrol at the Dominion Radio Astrophysical Observatory, Penticton, B.C.' *JRASC*, 59 (June 1965), 101–5
Covington, A.E., and W.J. Medd. 'Simultaneous Observations of Solar Radio Noise on 1.5 Meters and 10.7 Centimeters.' *JRASC*, 43 (May–June 1949), 106–10
Crommelin, A.C.D. 'The Astronomical Work of John S. Plaskett.' *JRASC*, 24 (May–June 1930), 217–32

Cumberland, J., and R.W. Cumberland. 'Nathan Fellowes Dupuis.' *Queen's Quarterly* (March 1927), 6ff

Daintry, Valentine. 'Investigation of the Rules Contained in Judge Fletcher's Paper.' *TLHSQ*, 4, part 1 (1843), 11–15

De Lury, A.T. 'The Eclipse Expedition to Labrador, August 1905.' *Transactions of the Toronto Astronomical Society*, 1905, 57–69

De Lury, Ralph E. 'The Effect of Haze on Spectroscopic Measures of the Solar Rotation: Explanation of Differences in Values, and Differences Depending on the Intensities of Spectrum Lines.' *JRASC*, 10 (Sept. 1916), 345–57

– 'The Law of the Solar Rotation.' *JRASC*, 33 (Nov. 1939), 345–78

– 'The Nature of a Suspected Variation of the Solar Rotation in 1915.' *JRASC*, 13 (Dec. 1919), 449–51

– 'The Question of the Presence of Haze Spectrum in the Mount Wilson Observations of the Solar Rotation.' *JRASC*, 11 (Jan. 1917), 23–4

– 'Sunspot Influences.' *JRASC*, 32 (March 1938), 105–32; (April 1938), 161–82

DeVecchi, V.G.M. 'Science and Government in Nineteenth-Century Canada.' PH D dissertation, University of Toronto, 1978

Deville, E.G.D. *Examples of Astronomic and Geodetic Calculations for the Use of Land Surveyors*. Quebec 1878

Doern, G. Bruce. *Science and Politics in Canada*. Montreal 1972

Douglas, A. Vibert. 'Astronomy at Queen's University.' *JRASC*, 52 (April 1958), 82–6

– *The Life of Arthur Stanley Eddington*. London 1956

– 'The St. Helena Observatory and Canadian Astronomy.' *Queen's Quarterly*, 78 (winter 1971), 592–601

– 'Spectroscopic Absolute Magnitudes and Parallaxes of 200 A-type Stars.' *JRASC*, 20 (Oct. 1926), 265–302

Douglas, James, jr. 'On Recent Spectroscopic Observations of the Sun, and the Total Eclipse of 7th August, 1869.' *TLHSQ*, n.s., part 7 (1869), 55–84

– 'Opening Address.' *TLHSQ*, n.s., part 4 (1866), 10–13

Drolet, Antonio. 'La bibliothèque du Collège des Jésuites.' *Revue d'histoire de l'Amérique française*, 14 (1960), 487–544

– *Les bibliothèques canadiennes, 1604–1960*. Ottawa 1965

– 'Ouvrages scientifiques.' *Naturaliste canadien*, 82 (April–May 1955), 105–6

Eddy, John. 'The Maunder Minimum.' *Science*, 192 (1976), 1189–1202

'Edouard Gaston Daniel Deville.' *PRSC* (1925), viii–xi

Eggleston, Wilfrid. *National Research in Canada*. Montreal 1978

Fletcher, John, 'On the Different Modes of Reducing the Apparent Distance between the Moon and the Sun, or a Star, in Lunar Observations, to the True Distance for Ascertaining the Longitude.' *TLHSQ*, 4, part 1 (1843), 1–10

Forbes, Eric G. *Greenwich Observatory: Origins and Early History (1675–1835)*. London 1975

Foster, J.S., and A.V. Douglas. 'Analysis of Profiles of Helium Lines in Spectra of B Stars.' *Nature*, 134 (1934), 417–18

– 'Stellar Stark Line He 4470.' *Physical Review*, 44 (1933), 325

Frégault, G., and M. Trudel, eds. *Histoire du Canada par les textes*. Montreal 1963

Gagnon, Ernest. *Louis Jolliet, découvreur du Mississippi*. Quebec 1902

Gaizauskas, V. 'The Ottawa River Solar Observatory.' *JRASC*, 70 (Feb. 1976), 1–22

Garneau, DeLisle. 'My Work with the Telescope.' *JRASC*, 33 (July–Aug. 1939), 251–4

Gosselin, Amédée. *L'instruction au Canada sous le régime français (1635–1760)*. Quebec 1911

Gosselin, Auguste. 'L'abbé Holmes et l'instruction publique.' *MSRC* (1907), sect. I, 127–72

– 'Les Jésuites au Canada: le Père de Bonnécamps, dernier professeur d'hydrographie au Collège de Québec (1741–1759).' *MSRC*, ser. 2 (1895), sec. I, 25–61; (1897), 93–118; (1898), 33–5

Halliday, Ian, A.T. Blackwell, and A. Griffin. 'The Innisfree Meteorite and the Canadian Camera Network.' *JRASC*, 72 (Feb. 1978), 15–39

Hamel, Thomas-Etienne. 'Science et ses ennemis.' *PRSC*, 5 (1887), xv–xxii

Hardin, Herschell. *A Nation Unaware: The Canadian Economic Culture*. Vancouver 1974

Harper, W.E. 'Atmospheric Conditions Suitable for the 72-inch Reflector.' *JRASC*, 8 (May–June 1914), 172–9

– 'The History of Astronomy in Canada.' *JRASC*, 32 (Oct. 1938), 381–90

Harris, Robin S. *A History of Higher Education in Canada, 1663–1960*. Toronto 1976

Harrower, George. 'Canadian Scientists Report XXI.' *JRASC*, 54 (Apr. 1960), 82–3

Heard, J.F. 'Clarence Augustus Chant.' *JRASC*, 51 (Feb. 1957), 1–4

– 'Ruth Josephine Northcott.' *JRASC*, 63 (Oct. 1969), 225–6

Heard, J.F., and H.S. Hogg. 'Astronomy at the David Dunlap Observatory 1935–1967.' *JRASC*, 61 (Oct. 1967), 257–76

Hearne, Samuel. *A Journey from Prince of Wales's Fort in Hudson's Bay to the Northern Ocean*. London 1795

Henning, Thomas. 'Meteors and Falling Stars.' *Canadian Journal*, 2 (1853–54), 188–91, 209–12

– 'Remarks on the Planetoids between Mars and Jupiter.' *Canadian Journal*, 3 (1854–5), 206–9

Henroteau, François. 'The Electronic Telescope.' *JRASC*, 28 (Feb. 1934), 59–62
- 'The Interstellar Clouds of Metallic Gases.' *JRASC*, 15 (Feb. 1921), 62–70; (May 1921), 109–19
Henroteau, F., and A.V. Douglas, 'A Study of Eta Aquilae.' *Pubs. DO*, 9, 7 (1929), 163–77
Henroteau, F., and H. Grouiller. 'International Co-operation for the Study of Cepheid Variables.' *JRASC*, 19 (Oct. 1925), 201–3
Henroteau F., and J.P. Henderson. 'A Spectroscopic Study of Early Class B Stars.' *Pubs. DO*, 5, 1 (1922)
Hey, J.S. *The Evolution of Radio Astronomy*. New York 1973
Hogg, Helen S. 'Early Days of Astronomy at Toronto.' *JRASC*, 75 (Dec. 1981), 281–8; 76 (Feb. 1982), 26–34; (June 1982), 149–56; (Aug. 1982), 235–44
- 'Globular Star Clusters.' *JRASC*, 52 (June 1958), 97–108
- 'John Frederick Heard, 1907–1976.' *JRASC*, 71 (Feb. 1977), 1–8
- 'Periods and Light Curves of the Variable Stars in the Globular Cluster Messier 2.' *Pubs. DAO* 6, 14 (1935), 265–84
- 'Some Contributions of R.M. Petrie to the Study of Groups of Stars.' *JRASC*, 61 (June 1967), 105–16
- 'Two Centenaries: The Royal Society of Canada and the Last Transit of Venus, 1882.' *JRASC*, 76 (Dec. 1982), 362–70
- 'Variable Stars in Globular Clusters.' *JRASC*, 67 (Feb. 1973), 8–18
Holland, Samuel. 'Astronomical Observations.' *Phil. Trans.*, 58 (1768), 247–52
- 'A Letter ... Containing Some Eclipses of Jupiter's Satellites, Observed near Quebec.' *Phil. Trans.*, 64 (1774), 171–6
- 'Observations Made on the Islands of Saint John and Cape Briton [sic].' *Phil. Trans*, 58 (1768), 46–53
James, Thomas. *The Dangerous Voyage of Capt. Thomas James, in His Intended Discovery of a North West Passage into the South Sea*. London 1633; reprinted 1740
Jarrell, Richard A. 'Astronomical Archives in Canada.' *Journal of the History of Astronomy*, 6 (1975), 143–7
- 'Astronomical Archives in Canada–2.' *Journal of the History of Astronomy*, 8 (1977), 71–2
- 'The Birth of Canadian Astrophysics; J.S. Plaskett at the Dominion Observatory.' *JRASC*, 71 (June 1977), 221–33
- 'British Scientific Institutions and Canada: The Rhetoric and the Reality.' *TRSC*, ser. 4, 20 (1982), 533–47
'Differential National Development and Colonial Science: The Cases of Ireland and Quebec.' In N. Reingold and M. Rothenburg, eds. *Scientific Colonialism: A Cross-Cultural Comparison, 1800–1930*. Washington, DC 1987
- 'Edward David Ashe.' *Dictionary of Canadian Biography*, XII (forthcoming)

- 'James Williamson.' *Dictionary of Canadian Biography*, XII (forthcoming)
- Origins of Canadian Government Astronomy.' *JRASC*, 69 (April 1975), 77–85
- The Reception of Einstein's Theory of Relativity in Canada.' *JRASC*, 73 (Dec. 1979), 358–69
- 'The Rise and Decline of Science at Quebec, 1824–1844.' *Histoire sociale / Social History*, 9 (May 1977), 77–91
- 'Science Education at the University of New Brunswick in the Nineteenth Century.' *Acadiensis* (spring 1973), 55–79

Kalm, Pehr. *Peter Kalm's Travels in North America*. Ed. A. Benson. 2 vols. New York 1966

Keeping, E.S. *A Short History of the Mathematics Department*, Edmonton 1971

Kemp, James. *The Centennial of the Toronto Centre*. Toronto 1968

Kennedy, J.E. 'The Brydone Jack Lectures on Astronomy and Related Topics.' *JRASC*, 75 (June 1981), 132–8
- 'The Development of Astronomy in Fredericton, New Brunswick, between 1847 and 1876.' *JRASC*, 70 (Oct. 1976), 238–46
- 'The Early Days of the First Astronomical Observatory in Canada.' *JRASC*, 49 (Sept.–Oct. 1955), 181–8
- 'An Early Exchange of Correspondence between King's and Queen's.' *Queen's Review* (Dec. 1954), 246–7, 267
- 'On the Solar Eclipse Expedition of 1860.' *JRASC*, 70 (April 1976), 74–6
- 'Our Heritage in Canadian Astronomy,' *JRASC*, 66 (April 1972), 83–98

King, Henry C. *The History of the Telescope*. London 1955

King, William F. 'The Coelostat House of the Dominion Observatory.' *JRASC*, 3 (Mar.–Apr. 1909), 115–16
- 'The Dominion Observatory at Ottawa.' *Transactions of the Toronto Astronomical Society* (1905), 27–34
- 'The New Reflecting Telescope for the Dominion Observatory.' *JRASC*, 7 (May–June 1913), 216–28

Kingston, H.R. 'The Hume Cronyn Memorial Observatory.' *JRASC*, 34 (Mar. 1940), 95–7

Klotz, Otto. 'The Dominion Astronomical Observatory at Ottawa,' *JRASC* 13 (Jan. 1919), 1–15
- 'Observatories in Canada.' *JRASC*, 12 (May–June 1918), 217–24; 13 (Sept. 1919), 322–32

Lacey, J.D. 'A Spectroscopic Supersynthesis Radio Telescope.' *JRASC*, 62 (Dec. 1968), 378–80

Lamontagne, Roland. *La Galissonière et le Canada*. Paris 1963
- *Roland-Michel Barrin de la Galissonière 1693–1756*. Quebec 1970

Lareau, Edmond. *Histoire de la littérature canadienne*. Montreal 1874

Le Clerq, Chrestien. *New Relation of Gaspesia*. Toronto 1910

Lefroy, John H. *In Search of the Magnetic North: A Soldier-Surveyor's Letters from the Northwest, 1843–1844.* Ed. G.F.G. Stanley. Toronto 1955

Levere, T.H., and R.A. Jarrell, eds. *A Curious Field-book: Science and Society in Canadian History.* Toronto 1974

Literary and Historical Society of Quebec. *The Centenary Volume of the Literary and Historical Society of Quebec.* Quebec 1924

Lortie, Léon. 'L'énseignement des sciences exactes en 1840 à Québec.' *Annales d'ACFAS,* 1 (1935), 92

Locke, J.L. 'Carlyle Smith Beals, 1899–1979.' *JRASC,* 73 (Dec. 1979), 325–32

– 'Recent Developments of Radio Astronomy in Canada.' *JRASC,* 61 (Oct. 1967), 324–38

– 'Report on Construction of the Canada–France–Hawaii Telescope.' *JRASC,* 71 (Feb. 1977), 9–20

Locke, J.L., J. Galt, and C. Costain. 'The 1420 MC/s Radio Telescope of the Dominion Radio Astrophysical Observatory.' *Pubs. DO,* 25, 7 (1965), 77–82

Lumsden, G.E. 'The Unification of Civil, Nautical, and Astronomical Time.' *TRSC,* ser. 2, 2 (1896), sec. III, 83–90

McDiarmid, F.A. 'Charles Albert Bigger.' *JRASC,* 17 (Nov.–Dec. 1923), 357–9

McDiarmid, R.J. 'Robert Millford Motherwell, 1882–1940.' *JRASC,* 34 (Dec. 1940), 441–3

McKellar, Andrew. 'Evidence for the Molecular Origin of Some Hitherto Unidentified Interstellar Lines.' *Publications of the Astronomical Society of the Pacific,* 52 (1940), 187–92

– 'Molecular Lines from the Lowest States of Diatomic Molecules Composed of Atoms Probably Present in Interstellar Space.' *Pubs. DAO,* 7, 15 (1941), 251–72

– 'Some Topics in Molecular Astronomy.' *JRASC,* 54 (June 1960), 97–109

Mackenzie Alexander. *Voyages from Montreal ... to the Frozen and Pacific Oceans in the Years 1789 and 1793.* London 1801

McKinley, D.W.R. 'Meteor Velocities Determined by Radio Observations.' *Ap. J.,* 113 (1951), 225–67

– *Meteor Science and Engineering.* New York 1961

McLeod, C.H. 'Sunspots Observed at McGill College Observatory.' *TRSC,* 8 (1890), sec. III, 43–7; 10 (1892), sec. III, 29–32

Magee, G.R. 'Harold Reynolds Kingston, 1886–1963.' *JRASC,* 57 (June 1963), 107–8

Marshall, J. Stewart. *Three McGill Weather Observatories.* Montreal 1968

Meadows, A.J. *Greenwich Observatory: Recent History (1836–1975).* London 1975

Meen, V.B. 'Chubb Crater, Ungava, Quebec.' *JRASC,* 44 (Sept.–Oct. 1950), 169–80

Middleton, W.E.K. *Physics at the National Research Council.* Waterloo 1980

Miller, Howard S. *Dollars for Research: Science and Its Patrons in 19th-Century America*. Seattle 1970

Millman, Peter M. 'The Meanook-Newbrook Meteor Observatories.' *JRASC*, 53 (Feb. 1959), 15–33

– 'Meteoric Ionization.' *JRASC*, 44 (Nov.–Dec. 1950), 209–20

– 'One Hundred and Fifteen Years of Meteor Spectroscopy.' *Solid Particles in the Solar System: IAU Symposium No. 90* ed. I. Halliday and B. McIntosh (1980), 121–8

– 'Reynold Kenneth Young, 1886–1977.' *JRASC*, 72 (Aug. 1978), 181–8

Millman, Peter M., and D.W.R. McKinley. 'Meteors.' *The Solar System*. Ed. B. Middlehurst and G. Kuiper. Vol. 4, 674–773, Chicago 1963

– 'A Note on Four Complex Meteor Radar Echoes.' *JRASC*, 42 (May–June 1948), 121–30

– 'Stars Fall over Canada.' *JRASC*, 61 (Oct. 1967), 277–94

Morgan, Henry J. *Bibliotheca Canadensis; or a Manual of Canadian Literature*. Ottawa 1867

Motherwell, R.M. 'The New Photographic Telescope of the Dominion Observatory.' *JRASC*, 8 (Sept.–Oct. 1914), 305–6

Mozel, Philip. 'The Woodstock College Observatory.' *JRASC*, 76 (June 1982), 168–80

Nadeau, Paul-H. 'Examen d'un miroir parabolique de 40 cm d'ouverture.' *JRASC*, 37 (July–Aug. 1943), 237–40

– 'The Quebec Centre of the Royal Astronomical Society of Canada.' *JRASC* 38 (Oct. 1944), 313–19

Northcott, Ruth J. 'The Growth of the RASC and its Guiding Mentor C.A. Chant.' *JRASC*, 61 (Oct. 1967), 218–25

Nugent, D.B. 'Charles Campbell Smith, 1872–1940.' *JRASC*, 34 (Nov. 1940), 405–8

Osterbrock, Donald E. 'The Rise and Fall of Edward S. Holden.' *Journal for the History of Astronomy* 15, 2 (June 1984), 81–127; 15, 3 (Oct. 1984), 151–76

Parker, T.H. 'John Beattie Cannon, M.A. 1879–1940.' *JRASC*, 35 (Jan. 1941), 15–19

Pearce, Joseph A. 'Report of the Dominion Astrophysical Observatory for the Years 1942–1945.' *JRASC*, 40 (April 1946), 139–55

– Report of the Dominion Astrophysical Observatory, Victoria, B.C. for the Years 1940 and 1941.' *JRASC*, 36 (July–Aug. 1942), 273–87

– 'Some Recollections of the Observatory (1924–1935).' *JRASC*, 62 (Dec. 1968), 287–97

Percy, John R. 'Graduate Programs in Astronomy at Canadian Universities.' *JRASC*, 71 (Jan. 1977), 264–74

– 'Undergraduate Astronomy at Canadian Universities.' *JRASC*, 72 (Oct. 1978), 296–8

Petrie, Jean K. 'R.M. Petrie and the B-Star Program of the Dominion Astro-physical Observatory.' *JRASC*, 64 (June 1970), 163–72

Petrie, Robert M. 'Absolute Magnitudes of the B-Stars Determined from Meas-ured Intensities of the H Lines.' *Pubs. DAO*, 9, 8 (1953), 251–67

– 'Absorption Line Intensities, Spectral Classification and Excitation Tempera-tures for the O Stars.' *Pubs. DAO*, 7, 21 (1947), 321

– 'The Growth of Astronomy in Canada.' *JRASC*, 50 (July–Aug. 1956), 146–51

– 'A Large Optical Telescope for Canada.' *JRASC*, 57 (Aug. 1963), 145–52

Petrie, Robert M., and J.A. Pearce. 'Radial Velocities of 570 B Stars.' *Pubs. DAO*, 12, 1 (1962), 1–90

Petrie, Robert M., and J.K. Petrie, 'Distribution and Motions of 688 B Stars.' *Pubs. DAO*, 13, 9 (1968), 253–72

Plaskett, Harry H. 'Atmospheric Haze and a Suspected Variation of the Solar Rotation in 1915.' *JRASC*, 13 (Nov. 1919), 391–402

– 'The Psychology of Differential Measurements.' *JRASC*, 10 (May–June 1916), 220–34

– 'The Solar Rotation in 1913.' *JRASC*, 8 (Sept.–Oct. 1914), 307–17

Plaskett, John S. 'Description of Building and Equipment.' *Pubs. DAO*, 1, 1 (1919), 7–103

– 'A Great Reflector for Canada.' *JRASC*, 7 (Nov.–Dec. 1913), 448–55

– 'The History of Astronomy in British Columbia.' *JRASC*, 77 (June 1983), 108–20

– 'The Motion of the Stars.' *JRASC*, 22 (April 1928), 111–34

– 'The O-Type Stars.' *Pubs. DAO*, 2, 16 (1924), 287–356

– 'The Rotation of the Galaxy.' *MNRAS*, 88 (March 1928), 395–403

– 'The 72-inch Reflecting Telescope.' *JRASC*, 8 (May–June 1914), 180–7

– 'The Solar Union.' *JRASC*, 7 (Nov.-Dec. 1913), 420–37

– 'The Spectroscopic Orbit of 56° 2617.' *Pubs. DAO*, 2, 14 (1923), 269–74

– 'The Star of Greatest Known Mass.' *JRASC*, 16 (Nov. 1922), 284–93

– 'W.F. King.' *JRASC*, 10 (July–Aug. 1916), 267–74

Plaskett, John S., and J.A. Pearce. 'A Catalogue of the Radial Velocities of O and B Type Stars.' *Pubs. DAO*, 5, 2 (1930), 99–165

– 'The Motions of the O and B Type Stars and the Scale of the Galaxy.' *Pubs. DAO*, 5, 4 (1934), 241–328

– 'The Motions and Distribution of Interstellar Matter.' *MNRAS*, 90 (Jan. 1930), 243–68

– 'The Problems of the Diffuse Matter in the Galaxy.' *Pubs. DAO*, 5, 3 (1931), 167–223

– 'The Radial Velocities of 523 O and B Type Stars Obtained at Victoria, 1923–1929.' *Pubs. DAO*, 5, 1 (1930), 1–98

Potter, A.F. *Catalogue of Optical, Mathematical and Philosophical Instruments and School Apparatus.* Toronto 1861

Provost, Honorius. *Historique de la faculté des arts de l'Université Laval, 1852–1952.* Quebec, 1952

– *Le Séminaire de Québec: documents et biographies.* Quebec 1964

'Queen's College Observatory.' *Queen's College Journal,* 26 (Nov. 1881), 1

Racine, René. 'L'astronomie au Québec.' *JRASC,* 70 (June 1976), 138–42

– 'L'Observatoire astronomique du Mont Mégantic : un nouvel observatoire d'envergure au Canada.' *JRASC,* 72 (Dec. 1978), 324–34

Raymond, W.O. *The Genesis of the University of New Brunswick with a Sketch of the Life of William Brydone Jack, A.M., D.C.L. President, 1861–1885.* Saint John 1919

Redman, R.O. 'The Galactic Rotation Effect in Late Type Stars.' *MNRAS,* 90 (May 1930), 690–6

– 'The Galactic Rotation Effect in Some Late Type Stars.' *Pubs. DAO,* 4, 20 (1930), 325–40; 6, 5 (1931), 27–48

Report Submitted to the Space Science Coordination Office, National Research Council. Ottawa 1977

Richardson, E.H. 'The Spectrographs of the Dominion Astrophysical Observatory.' *JRASC,* 62 (Dec. 1968), 313–30

Richardson, E.H., G. Brealey, and R. Dancey. 'An Efficient Coudé Spectrograph System.' *Pubs. DAO,* 14, 1 (1971), 1–15

Robb, James. *Oration Delivered at the Encaenia in King's College, Fredericton, June 27, 1839.* Fredericton 1839

Roeder, R.C., and P.P. Kronberg. 'Canadian Astronomy: Manpower Supply and Demand.' *JRASC,* 64 (Oct. 1970), 315–18

Rogers, R.V. 'Professor James Williamson, LL.D.' *Queen's Quarterly,* 3 (1895), 161–72

Rogers, W.A., and C.H. McLeod. 'The Longitude of McGill College Observatory.' *TRSC,* 3 (1885), sec. III, III–37

Ross, Sir John. *Narrative of a Second Voyage in Search of a North-West Passage.* London 1835

– *A Voyage of Discovery ... for the Purpose of Exploring Baffin's Bay and Enquiring into the Probability of a North-West Passage.* London 1819

Rottenburg, Col Baron de. 'On an Occultation of Spica Virginis by the Moon.' *Canadian Journal,* n.s. 2 (1857), 180–5

– 'Solar Spots Observed at Toronto in January, February, and March, 1858.' *Canadian Journal,* n.s. 3 (1858), 293–7

– 'The Supposed Self-Luminosity of the Planet Neptune.' *Canadian Journal,* n.s. 1 (1856), 424–9

Roy, Antoine. *Les lettres, les sciences et les arts au Canada sous le régime français.* Paris 1930

Roy, P.-G. 'Un hydrographe du roi à Québec, Jean-Baptiste Louis Franquelin.' *MSRC*, Ser. 3, 13 (1919), 47ff

Royal Society of Canada. *The Royal Society of Canada 1882–1957.* Ottawa 1958

Ryerson, Egerton. *On the Course of Collegiate Education.* Toronto 1842

St. John, C.E., and W.S. Adams. 'The Question of Diffused Light in Mount Wilson Solar Observations.' *JRASC*, 10 (Dec. 1916), 553–5

Scadding, Henry. *The Astrolabes of Samuel Champlain and Geoffrey Chaucer.* Toronto 1880

Segal, B. *A Preliminary Guide to the C. Kirkland McLeod Collection of Clement Henry McLeod Papers.* Montreal 1972

Smallwood, Charles. 'The Observatory at St. Martin, Isle Jesus, Canada East.' *Canadian Journal*, n.s. 3 (1858), 281–92

Smith, Arthur. 'L'Observatoire de Québec.' *Bulletin de Recherche Historique*, 42 (1936), 16–18

Stewart, R.M. 'Dr. Otto Klotz.' *JRASC*, 18 (Jan.–Feb. 1924), 1–8

– 'The Early History of Astronomical Activity in the Canadian Public Service.' *JRASC*, 65 (Oct. 1971), 206–16

– 'William Edmund Harper, 1878–1940.' *JRASC*, 34 (July–Aug. 1940), 233–7

Struve, Otto, and V. Zebergs. *Astronomy of the Twentieth Century*, New York 1962

Swings, P. 'The Interstellar Absorption Lines of Molecular Origin.' *JRASC*, 35 (Feb. 1941), 71–80

Taylor, Henry. *A System of the Creation of Our Globe, of the Planets, and the Sun of Our System.* 9th edition. Quebec 1854

Thiessen, A.D. 'The Founding of the Toronto Magnetic Observatory and the Canadian Meteorological Service.' *JRASC*, 34 (Sept. 1940), 308–48

– 'Her Majesty's Magnetical and Meteorological Observatory, Toronto.' *JRASC*, 35 (April 1941), 141–50, 205–24; 36 (Feb. 1942), 61–5; (Dec. 1942), 457–72; 39 (July–Aug. 1945), 221–30; (Sept. 1945), 267–78; (Oct. 1945), 311–19; (Nov. 1945), 355–69; (Dec. 1945), 394–408

Thomson, Don W. *Men and Meridians.* 3 vols. Ottawa 1966–9

Thomson, Malcolm M. 'Astronomy of Time and Position.' *JRASC*, 61 (Oct. 1967), 226–40

– *The Beginning of the Long Dash: A History of Timekeeping in Canada.* Toronto 1978

Thwaites, R.G. ed. *The Jesuit Relations and Allied Documents.* 73 vols. New York 1959

Turnor, Philip. *Results of Astronomical Observations made in the Interior Parts of North America.* London 1794

Underhill, Anne B. 'The Interpretation of Stellar Spectra.' *JRASC*, 62 (Dec. 1968), 331–43

Underhill, Anne B., and William Petrie. 'The Stark Effect of Helium in Some B-type Stars.' *JRASC*, 38 (Nov. 1944), 385–94

Wales, William, and Joseph Dymond. 'Astronomical Observations Made by Order of the Royal Society at Prince of Wales's Fort on the North-west Coast of Hudson's Bay.' *Phil. Trans.*, 59 (1769), 467–88

Walker, G.A.H. 'Studies of the Interstellar Medium at the Dominion Astrophysical Observatory.' *JRASC*, 62 (Dec. 1968), 361–6

Wallace, R.C., ed. *Some Great Men of Queen's.* Toronto 1941

Wallace, W.S. *A History of the University of Toronto 1827–1927.* Toronto 1927

– *Royal Canadian Institute Centennial Volume, 1849–1949.* Toronto 1949

Watson, Albert D. 'Astronomy in Canada.' *JRASC*, 11 (Feb. 1917), 47–78

Wehlau, William. 'New Observatory of the University of Western Ontario.' *JRASC*, 64 (Feb. 1970), 1–4

Welsh, H.L. 'Gerhard Herzberg – Nobel Laureate, 1971.' *JRASC*, 66 (Aug. 1972), 183–8

Whitfield, Carol, and Richard A. Jarrell. 'John Henry Lefroy.' *Dictionary of Canadian Biography.* Vol. XI, 508–10. Toronto 1982

Wiggins, Ezekiel Stone. *The Architecture of the Heavens, Containing a New Theory of the Universe and the Extent of the Deluge ... in Opposition to the views of Dr. Colenso.* Montreal 1864

Williamson, Isabel K. 'The Observatory of the Montreal Centre.' *JRASC*, 52 (Feb. 1958), 3–4

Williamson, James. 'Determination of the Latitude of Kingston Observatory, Canada.' *MNRAS*, 28 (Nov. 1867), 12–13

– 'Longitude of Kingston.' *Canadian Journal*, 3 (1854–5), 82

Woolsey, E.G. 'Catalogue Contributions of the Ottawa Meridian Circle.' *JRASC*, 58 (Apr. 1964), 68–78

– 'The Ottawa Meridian Circle.' *JRASC*, 53 (Dec. 1959), 264–70

Working Group on Space Astronomy. *A Report Submitted to the Ad Hoc Committee on Canadian Scientific Experiments for the Space Shuttle / Spacelab Program.* Ottawa 1977

Wright, K.O. 'The Dominion Astrophysical Observatory 1918–1975.' *JRASC*, 69 (Oct. 1975), 205–11

– 'Fifty Years at the Dominion Astrophysical Observatory.' *JRASC*, 62 (Dec. 1968), 269–86

– 'Stellar Atmospheres and Their Spectra.' *JRASC*, 60 (June 1966), 97–118

Wright, K.O. et al. 'The Future of Optical Astronomy in Canada.' *JRASC*, 57 (April 1963), 66–72

Young, R.K. 'The Building of a 19-inch Reflecting Telescope.' *JRASC*, 24 (Jan. 1930), 17–33

– 'The Calcium Lines H and K in Early Type Stars.' *Pubs. DAO*, 1, 17 (1920), 219–31

– 'The Calcium Lines H and K in Early Type Stars.' *JRASC*, 14 (Dec. 1920), 389–408

– 'The David Dunlap Observatory.' *University of Toronto Quarterly*, 4 (1935), 327–36

– 'David Dunlap Observatory. Outline of Work, July 1940–July 1941.' *JRASC*, 36 (April 1942), 142–4

– 'The 74-Inch Telescope of the David Dunlap Observatory.' *JRASC*, 28 (March 1934), 97–119

– 'The Spectroscopic Binary 12 Lacertae.' *Pubs. DAO*, 1, 2 (1918), 105–11

Young, R.K., and W.E. Harper. 'The Absolute Magnitudes and Parallaxes of 1080 Stars.' *JRASC*, 18 (Jan.–Feb. 1924), 9–59

Index

Picture Credits

The photograph on the jacket was taken by D'Arcy R.G. Jarrell. The illustrations in the text (appearing before pages 29, 87, and 153) appear with the kind permission of the following people and institutions:

Royal Astronomical Society of Canada (RASC), Archives: Jack (photo by J.E. Kennedy); eclipse expedition, 1869; Toronto-area amateurs; Montreal Centre, RASC; Douglas; Thomson

National Archives of Canada (NAC): International Boundary Commission, 1893–5, PA-12444; Darrah, C-78982; Klotz, PA-12295; McGill Observatory, C-8311; Dominion Observatory (Topley), PA-10296; portable transit instrument, PA-107537; DO eclipse expedition, 1905, PA-138911; opening of DAO, PA-149323; coelostat house, DO, PA-107522; De Lury (J.B. Scott/NFB), PA-166847; Stewart (J.B. Scott/NFB), PA-166848; Petrie (Garnet Lunney/NFB), PA-166845; Herzberg, PA-128062; feed horn, DRAO (Chris and Gar Lunney Lund/NFB), PA-166843

Notman Photographic Archives, McCord Museum, McGill University: Quebec Observatory and time ball

Metropolitan Toronto Library: Second Toronto Observatory

University of Toronto Archives: DDO, 3 June 1938, A65 0004/011; DDO 74-inch telescope, A78 0041/027; Helen Hogg on platform, A78 0041/027; DDO staff, 1964, A78 0041/027; astronomy class, A78 0041/027; 46-m telescope at Algonquin, A78 0041/027

Peter M. Millman: meteor observers

Helen S. Hogg: Helen Hogg and Frank Hogg

Lightning Source UK Ltd.
Milton Keynes UK
UKHW010004210722
406167UK00001B/137

9 781487 592028